G. Lister Sutcliffe

The Principles and Practice of Modern House-Construction

G. Lister Sutcliffe

The Principles and Practice of Modern House-Construction

ISBN/EAN: 9783744666527

Printed in Europe, USA, Canada, Australia, Japan

Cover: Foto ©berggeist007 / pixelio.de

More available books at **www.hansebooks.com**

THE PRINCIPLES AND PRACTICE
OF
MODERN HOUSE-CONSTRUCTION

INCLUDING

WATER-SUPPLY AND FITTINGS—SANITARY FITTINGS AND
PLUMBING—DRAINAGE AND SEWAGE-DISPOSAL—WARMING
VENTILATION — LIGHTING — SANITARY ASPECTS OF FUR-
NITURE AND DECORATION — CLIMATE AND SITUATION
STABLES—SANITARY LAW, &c.

WRITTEN BY

F. W. ANDREWES, M.D., F.R.C.P., D.P.H.
A. WYNTER BLYTH, M.R.C.S., F.I.C.
H. PERCY BOULNOIS, M.Inst.C.E., F.San.I.
E. A. CLAREMONT, M.I.E.E., M.I.M.E.
HENRY CLAY, R.I.P.L.
E. R. DOLBY, A.M.Inst.C.E., M.I.M.E.
WILLIAM HENMAN, F.R.I.B.A.
H. JOSSÉ JOHNSON, M.B., D.P.H.

Prof. ROBERT KERR, F.R.I.B.A.
HENRY LAW, M.Inst.C.E., F.San.I.
F. W. LOCKWOOD, F.I.S.E.
J. MURRAY SOMERVILLE
W. SPINKS, A.M.Inst.C.E., Pr.I.S.E.
G. LISTER SUTCLIFFE, A.R.I.B.A., M.San.I.
WILLIAM H. WELLS
E. F. WILLOUGHBY, M.D., D.P.H.

KEITH D. YOUNG, F.R.I.B.A.

EDITED BY
G. LISTER SUTCLIFFE

ARCHITECT
ASSOCIATE OF THE ROYAL INSTITUTE OF BRITISH ARCHITECTS, MEMBER OF THE SANITARY INSTITUTE
AUTHOR OF "CONCRETE: ITS NATURE AND USES", ETC.

ILLUSTRATED BY ABOVE 700 FIGURES IN THE TEXT, AND A SERIES OF SEPARATELY-PRINTED PLATES

DIVISIONAL-VOL. V.

LONDON: BLACKIE & SON, LIMITED
GLASGOW AND DUBLIN

With regard to the Illustrations in Divisional-Volume V., indebtedness has to be acknowledged to the following firms:—Messrs. Thos. Ash & Co., Birmingham; Jas Bedford & Co., Ltd., Halifax; Wm. Bruce, Hyde Park, Leeds; Donald & Donald, Glasgow; "Grahtryx" Ventilating and Engineering Co., Ltd., London; Jas. Howorth & Co., Farnworth, near Bolton; D. Bruce Peebles & Co., Bonnington, Edinburgh; The Premier Lighting Co., Salford; Renton Gibbs & Co., Ltd., Liverpool; F. H. Royce & Co., Ltd., Manchester; Wm. Sugg & Co., Ltd., London; P. M. Walker & Co., Halifax; The Welsbach Incandescent Gas Light Co., Ltd., London.

CONTENTS.

DIVISIONAL-VOL. V.

SECTION XII.—VENTILATION.
By WILLIAM HENMAN, F.R.I.B.A.

                                                                                                             Page

CHAP. I.—GENERAL FACTS AND PRINCIPLES. The object of the section—Unscientific terms—What ventilation is—Physical properties of air—Contamination of air—Effects of respiration—Composition of air—Effects of contaminated air—Natural contamination—Contamination by human agencies—Natural processes of purification—Purification of air in buildings—Complex nature of the subject—Duties of the State and Local Authorities—Duties of individuals—Necessity for ventilation · 167

CHAP. II.—NATURAL FORCES MAKING FOR VENTILATION. Inlets and outlets essential—Effects of opening a window—Deductions therefrom—Mechanical power—Natural forces—Temperature—Movements of external air—Methods of warming—Circulation of air in houses—Up-draught in flues—Syphonic action—Down-draughts in flues—The necessity for air-inlets—Draughts—The sizes, nature, and positions of inlets and outlets—Velocity of air-currents—Some simple experiments · 174

CHAP. III.—THE PLANNING AND CONSTRUCTION OF HOUSES WITH REGARD TO THEIR VENTILATION. Aspect and situation—Windows—Bay-windows—Detached houses—Houses in rows or terraces—Courts and alleys—Back-to-back houses—Lofty houses—Impervious materials for walls—Partitions and floors—Examination of house for ventilation, draughts, air-pollution, &c.—Roof-spaces and roofs · · · · · · · · · · · · · 184

CHAP. IV.—AIR-CURRENTS AND AIR-INLETS. Detection of air-currents in rooms—Kitchen-smells—Tobin tubes—Openings with flaps or louvres—Size of inlets—Trumpet-mouthed inlets—The Sheringham inlet—Inlets with baffle-plates, water-drawers, &c.—Drawer inlets—Louvred inlets—Air-inlets in windows—Air-propellers or fans—Air-inlet cowls—Warmed-air inlets · · · · 190

CHAP. V.—AIR-OUTLETS. Smoke-flues—Mica-flap outlets—Improved mica-flap outlets—Outlets in external walls—Extract-cowls—Open tubes with protecting caps—Walker's extract-cowl—Cowls with movable parts—Lobster-back rotary cowl—Rotary ventilator for window—Howorth's Archimedean screw ventilator—Cowls without movable parts—Donald & Sime's—Kite's—Bedford's—Renton Gibbs's—Walker's—Sugg's—Kite's dove-cot extract-ventilator—Baird, Thompson, & Co.'s dove-cot extract-ventilator—Donald & Sime's concealed roof ventilator—Kite's chimney-stack ventilator—Water-spray ventilators · · · · 196

CHAP. VI.—"NATURAL" VENTILATION. Upward and downward ventilation—Fireplace-flues—Registers—Regulation of air-inlets—Windows and doors—Position of doors with regard to draughts—Ventilation of rooms through hall and stairway—Intelligent regulation necessary 20

CHAP. VII.—CONTAMINATION OF AIR, &c. Principal causes of contamination—Deposits of organic matter—Condensation—Evaporation—Absorption—Draperies—Dust—Ground-air—Basement-rooms—Spaces under floors—Foul air from drains and waste-pipes—Faulty construction—Importance of daylight—Artificial illumination by candles, oils, and gas—Ventilating gas-lights—Electricity—Contamination of surfaces in rooms—Disinfectants 208

CHAP. VIII.—WARMING AND VENTILATION. Various methods of warming—Ventilating fire-grates—Ordinary grates—Closed stoves—Gas-fires—Bruce's single and double apparatus for warming incoming air by gas—Bruce's hot-water ventilating radiator heated by gas—Arrangement and construction of buildings with regard to warming and ventilation—Hot-water heating—Objections to pipe-trenches—Steam heating—Ventilating radiators—Hot air—Air-ducts—Exits 214

CHAP. IX.—SUMMARY. General principles—"Natural" ventilation requires constant attention—Mechanical ventilation—Downward ventilation—Popular fallacies—Conclusion 222

SUPPLEMENTARY CHAPTER BY THE EDITOR.

VENTILATION BY MEANS OF WARMED AIR. Examples of houses designed with strict regard to ventilation—Drs. Drysdale and Hayward's writings and houses—Mr. E. J. May's designs—Fresh-air chamber and ducts—Extract-flues—Windows hermetically closed and otherwise—Objections to combined systems of warm-air supply and ventilation—Mr. May's improved arrangements—Reasons for tardy adoption of such systems—Mr. Bruce's apparatus for warming air by gas—Purification and humidification of air 225

SECTION XIII.—LIGHTING.

PART I.—CANDLES, OILS, AND ELECTRICITY.

By E. A. CLAREMONT, M.I.E.E., M.I.M.E.

CHAP. I.—CANDLES AND OILS.—The illuminating power of candles—"Candle-power" defined—Mineral burning oils—"Flashing-point" and "burning-point"—Russian and American oils—The choice of wicks—Dangers of oil-lamps—Features of good lamps—Comparison of the cost of illuminants—General comparison of illuminants—Various kinds of oil-lamps—Methods of suspending and elevating lamps . . . 231

CHAP. II.—ELECTRICITY: DEFINITIONS. Volts, voltage, electro-motive force—Ampère and current—Watts and electrical energy—Ohm—Positive and negative . 236

CHAP. III.—ELECTRICITY: GENERATION AND STORAGE. The primary battery—The dynamo—Magnets and electro-magnets—Armatures—Commutators—Starting a dynamo—1. The "Series" dynamo—2. The "Shunt" dynamo—Comparison of series and shunt dynamos—The limit of current in wires—"Short-circuit" defined—3. The "Compound" dynamo—Resistance-frames—The uses of the three types of dynamo—The polarity of a dynamo—Dynamos joined in series—Dynamos for private electric-light installations—Electric power-transmission—

Alternating-current dynamos—Commutators—Continuous and alternating current dynamos compared—Transformers—Wires—Alternating-current transformers—Interrupters—Continuous-current transformers—Motive powers for dynamos—Accumulators—Accumulator-cells in "series", and in "parallel" or "multiple"—Plates in cells—Selection of cells—Cell-connections—Capacity of cells—Charging the cells—Methods of preventing "spraying"—Hand-voltmeters—How to find the total voltage of cells—Ammeters 237

CHAP. IV.—ELECTRICITY: WIRING AND LAMPS. Fundamental principle of wiring—Size of wires—Voltage for lighting—Candle-power of lamps—Efficiency of lamps, "light-giving" and "life"—Improper treatment of lamps—Connection of lamps with circuit—The "tree" system of wiring—Switches—Double-pole switches—Lamp-holders—Switch lamp-holders—Short circuits, their dangers and prevention—Fusible cut-outs—Ceiling-roses—Wall-sockets—Portable lamps, &c.—The atmosphere an insulator—Trigger-action switches—Insulated copper wire—Quality of insulation—Insulated joints—The "distributing" system of wiring—Compromise between the two systems of wiring—Wood casing for wires—Lighting by arc-lamps—The carbon points—Arc-lamp resistances—The light from arc-lamps—Movement of carbons—Arc-lamps with reflectors and shades—Modifications of arc-lamps . 258

CHAP. V.—ELECTRICITY: AN INSTALLATION IN A HOUSE. The lighting of a country house—Preliminary calculations—The motive power—Selection of dynamo—The accumulator shelves—Foundations for engine and dynamo—The switchboard—The water-circulating tanks—The mains—The distributing boards—Lamps and fittings—Switches—Wiring—The exhaust from the engine—The belt for the dynamo—Charging the cells—Automatic accumulator-switches—Cut-outs—Printed instructions for working the installation 272

PART II.—GAS.
By HENRY CLAY.

CHAP. I.—METERS. The government stamp on meters—Faults of meters—"Wet" and "dry" meters compared—Wet meters—Dry meters—Frost and wet meters—Inlets and outlets of meters for various numbers of lights—Economy in lighting—Connections with meters—Stamping fees—The durability of meters 279

CHAP. II.—GAS-PIPES AND FITTINGS, &c. Lead pipes—Composition pipes—Iron pipes—Main service-pipes—Stop-cocks—Sizes of pipes—Fixing gas-pipes—Syphon-boxes—Syphons—Connections with wet and dry meters—Gas-taps—Gas-governors—Bruce, Peebles, & Co.'s mercurial gas-governor—The Wenham, Stott, and Shaw gas-governors—Bruce, Peebles, & Co.'s "Needle" governor-burner, and "Automatic" governor for stoves and fires—By-pass taps—Brass unions—Blocks . . 283

CHAP. III.—GAS-BURNERS. Bray's burners—Consumption of gas—The albo-carbon burner—Argand burners—Regenerative burners—The Wenham lamp and ventilating lamp—Incandescent gas-burners—The Welsbach burners—Comparison of incandescent burners—Cost and maintenance—Illuminating power—Vitiation of air—The Welsbach "C" burner—"C" burner with governor—"C" burner with by-pass—"C" burner with by-pass and governor—"S" burner with by-pass—Incandescent ventilating lamps—Reflectors—Burners for oil-gas . . . 289

CHAP. IV.—GAS-LEAKAGES AND EXPLOSIONS. Faulty gas-taps—Loose brackets and pendants—Leakages from water-slide pendants—Leakages from pipes—Periodical testing necessary—How to deal with an escape of gas 299

SECTION XIV.—GAS-PRODUCING APPARATUS FOR THE ILLUMINATION OF COUNTRY HOUSES.

By J. MURRAY SOMERVILLE.

Page

CHAP. I.—COAL-GAS. Four kinds of illuminating gas—Coal for gas-making—The retort-house and coal-store—Retorts—Ascension-pipes—Hydraulic mains—Condensers—The work in the retort-house—Deposits of tarry substances—Quantity of "liquor" in the hydraulic main—Quenching the ashes—Withdrawing the charges—Storage of coal—The work of the condensers—The tar and its uses—The two "nuisances"—The scrubber—Purifiers—The object of purification—The gas-meter and gas-holder—Gas-governors—The advantages of coal-gas 305

CHAP. II. OIL-GAS. Two methods of manufacture—Pintsch's process—General arrangements—The retorts—The working of the apparatus—Tar—The condensers—Purifiers and gas-holders—The illuminating power of oil-gas—Quantity of gas produced—Heating and cooking stoves for oil-gas—The Peebles process of manufacture—Cost of oil-gas 312

CHAP. III.—ACETYLENE. Its composition and illuminating power—Carbide of calcium—Acetylene generators—The light from acetylene—Burners—Advantages of acetylene for illuminating country houses—Essential conditions—The dangers of acetylene—Orders in Council as to storage of carbide of calcium and acetylene—Cost of acetylene 316

CHAP. IV. SPIRIT-GAS. Suitable spirits for converting into gas—The apparatus required—Quality and cost of spirit-gas—Advantages of spirit-gas for illumination - 319

SECTION XV.—THE SANITARY ASPECT OF DECORATION AND FURNITURE.

By EDWARD F. WILLOUGHBY, M.D., D.P.H., &c.

1.—GENERAL CONSIDERATIONS. Health, comfort, and pleasure—Light—Effect of colours on the senses—Dirt—Moisture—Pervious and impervious wall-surfaces . 323

2.—WALLS AND CEILINGS. Limewashing—Whitewashing—Distemper—Washable distempers—Water-glass—Frescoes—Oil-paints—Wall-papers—Arsenic in wall-papers, cretonnes, &c.—Washable substitutes for wall-papers—Ceilings . . 325

3.—BLINDS AND CURTAINS. Curtains—Blinds 329

4.—WOODWORK. Paint—Polish—Varnish 330

5.—FLOORS AND FLOOR-COVERINGS. Faulty floor-boards—Carpets—Varnished and wax-polished floors—Carpet squares—Linoleum, cork-carpet, &c. . . . 330

6.—FURNITURE. Improvement in taste—General principles of furniture design . 331

7.—SWEEPING, DUSTING, &c. Dust—Sweeping—Dusting—Use of damp cloths—Furniture-polish—Cleaning oil-paintings—Protecting the leather bindings of books 332

8.—PLANTS IN ROOMS. The "respiration" of plants—Effect of gas on plants—"Wardian" cases—Ferneries, conservatories, &c. 333

ILLUSTRATIONS.

DIVISIONAL-VOL. V.

		Page
Plate XX.	Plans and View of House at Chiswick Heated and Ventilated by Means of Warmed Air	225
,, XXI.	Section and Details of House at Chiswick, showing Warming and Ventilating Apparatus	226
,, XXII.	Plans and View of Two Houses at Hampstead, one of which is Heated and Ventilated by means of Warm Air and Open Fire-grates, &c.	227
,, XXIII.	Bruce's Warming and Ventilating Apparatus, Heated by Gas	228
,, XXIV.	Elevation and Plan of Gas-producing Plant suitable for Country Houses	305

LIST OF ILLUSTRATIONS IN TEXT.

SECTION XII.

Fig.		Page
561.	Section of House, showing Danger of Down-draughts in Flues	179
562.	Section of House with Extract-ventilator, showing Danger of Down-draughts	180
563.	Plan and Section showing Air-current in Room, with Square Inlet opposite the Outlet	181
564.	Plan and Section showing Air-current in Room, with Trumpet-mouthed Inlet on the same side as the Outlet	181
565.	View of Tobin Air-inlet Tube	190
566.	Short Tobin Tube	191
567.	Trumpet-mouthed Air-inlet	192
568.	Sheringham Air-inlet	192
569.	Air-inlet with Perforated Inlet-plate	193
570.	Sheringham Air-inlet with Baffle-plates	193
571.	Sheringham Air-inlet, with Baffle-plates and Inner Wind-guard	193
572.	Sheringham Air-inlet in Wood Bracket	193
573.	Air-inlet with Regulating Valve, and Brass Gauze Front	194
574.	Air-inlet with Baffle-plate, Valve, and Water-drawer	194
575.	Drawer Air-inlet with Baffle-plates	194
576.	Air-inlet with Movable Louvres	194
577.	Air-inlet for fixing on Bottom Rail of Window	194

Fig.		Page
578.	Howorth's Two-bladed Air-propeller or Fan	195
579.	Baird, Thompson, & Co.'s Air-inlet Cowl	195
580.	Elevation and Section of Donald & Sime's Air-inlet Cowl	195
581.	View of Baird, Thompson, & Co.'s Air-inlet Cowl	195
582.	Mica-flap Air-outlet	196
583.	Section of Mica-flap Air-outlet with Double Flaps	196
584.	Improved Mica-flap Air-outlet	197
585.	Projecting Air-outlet for fixing in Walls	197
586.	Flush Air-outlet for fixing in Walls	197
587.	Walker's Extract-cowl with Two Shafts	198
588.	View of Lobster-back Rotary Cowl	199
589.	Rotary Ventilator fixed in Window-pane	199
590.	Howorth's Archimedean Screw Revolving Ventilator	199
591.	Donald & Sime's Extract-cowl	200
592.	Kite's Extract-cowl in Wood Turret	200
593.	Bedford's Exhaust-ventilator	201
594.	Donald & Sime's Extract-cowl	201
595.	Benton Gibbs's Extract-cowl	202
596.	Walker's Extract-cowl	202
597.	Sugg's Extract-cowl	203
598.	Kite's Extract-ventilator of the Dove-cot Type	203

Fig.		Page
599.—Baird, Thompson, & Co.'s Extract-ventilator		202
600.—Donald & Sime's Concealed Roof-ventilator		203
601.—Kite's Chimney-stack Ventilator		204
602.—Kite's Water-spray Vertical Exhaust, driving downwards		204
603.—Four Arrangements of Doors and Fireplaces, showing Currents of Air		207
604.—Bruce's Single Apparatus for Warming Incoming Air by Gas		216
605.—Bruce's Double Apparatus for Warming Incoming Air by Gas		217
606.—Bruce's Hot-water Apparatus or Radiator heated by Gas		218
607.—Building showing Bruce's Radiators Warming the Incoming Air, and Upcast-shafts		219
608.—Building showing Bruce's Central Air-warmer and Shafts		220
609.—Steam Battery or Radiator admitting Fresh Warmed Air		221

SECTION XIII.

Fig.		Page
610.—Magnet and Lines of Force		238
611.—Electro-magnet		238
612.—Shaft with Discs forming Armature		239
613.—Armature partially Wound		239
614.—Armature and Commutator		239
615.—Diagram showing Series Dynamo and Circuit		240
616.—Diagram showing Shunt Dynamo and Circuit		240
617.—Diagram showing Compound Dynamo and Circuit		243
618.—Shunt Resistance-frame		243
619.—View of Transformer		249
620.—View of Accumulator-cell		251
621.—Accumulator-cells connected in Series		252
622.—Hand-voltmeter and Spear		256
623.—Diagram of Tree-wiring		261
624.—View of Small Switch		262
625.—View of Small Switch with Outer Case Removed		262
626.—View of Double-pole Main Switch		262
627.—Section of Lamp-holder		263
628.—View of Lamp-holder with Outer Case		263
629.—View of Switch Lamp-holder		263
630.—A Fusible Cut-out		264

Fig.		Page
631.—A Ceiling-rose		265
632.—Wall-socket and Plug		265
633.—Portable Hand-lantern		266
634.—Diagram of Distributing System of Wiring		268
635.—Section of Wood Case and Lid for Electric Wires		268
636.—An Arc-lamp		269
637.—Arc-lamp Resistance		270
638.—Arc-lamp with Half-globe Reflector		271
639.—Inverted Arc-lamp		271
640.—Sectional Elevations of Wet Gas-meter		280
641.—Sectional Elevation of Dry Gas-meter		281
642.—Connection of Internal Gas-main to Dry Meter		285
643.—Section of Bruce, Peebles, & Co.'s Mercurial Gas-governor		286
644.—Wenham Gas-governor		287
645.—Stott Gas-governor		287
646.—Shaw Gas-governor		287
647.—Bruce, Peebles, & Co.'s Needle Governor-burner		288
648.—Bruce, Peebles, & Co.'s Automatic Gas-governor for Stoves and Fires		288
649.—Three Bray's Gas-burners		289
650.—Sugg's Argand Burner		291
651.—Elevation of Wenham Lamp		291
652.—Wenham Ventilating Lamp		292
653.—Vertical Section through the Wenham Burner		292
654.—Welsbach Incandescent "C" Burner		295
655.—Parts of the Welsbach Incandescent "C" Burner		295
656.—Broach for Regulating the Burner		295
657.—Welsbach Incandescent "C" Burner with By-pass		296
658.—Three-light Incandescent Ventilating Lamp		297
659.—Large Incandescent Ventilating Lamp		298

SECTION XIV.

Fig.		Page
660.—Longitudinal Section of Retort		306
661.—Section of Hydraulic Main		307
662.—Section of Purifier		309
663.—Ground-floor Plan of Works for Pintsch's Oil-gas Process		312
664.—Goose-neck Seal for Running Oil into Retort		313
665.—Acetylene-gas Generator		317

Section XII.—Ventilation

BY

WILLIAM HENMAN

ARCHITECT

FELLOW OF THE ROYAL INSTITUTE OF BRITISH ARCHITECTS; PAST PRESIDENT OF THE BIRMINGHAM
ARCHITECTURAL ASSOCIATION

SECTION XII.—VENTILATION.

CHAPTER I.

GENERAL FACTS AND PRINCIPLES.

In dealing with the subject of ventilation in any form, it is necessary to touch upon elementary facts, which are applicable to whatever class of building or place it may be required to ventilate, yet anything like a detailed description of the special and more complicated methods which may be required for public buildings and institutions, or for factories, is beyond the scope of this work, and will not be attempted. **The object of this section of the book** is to place before architects, builders, students, and householders, such a view of the subject as will enable them intelligently to comprehend the possibilities of securing efficient ventilation in our British homes; for, although the principles are the same the whole world over, there are in some regions climatic conditions and national customs which require distinctive consideration and treatment. It is, however, undesirable herein to complicate the subject by discussing such special and somewhat exceptional demands.

Important as it may be that every architect and builder should have **a clear conception of what can be done** to provide for efficient ventilation throughout a dwelling, and that they should apply such knowledge in a practical manner, it is even more essential that every householder should understand the limits to the power of an architect or builder to make provision for ventilation under varying conditions of the atmosphere, and in the various apartments of a dwelling, and that he should also realize the necessity for constant supervision and regulation, even when the greatest care and knowledge have been employed in providing suitable means for ventilation.

Nothing has been more fruitful in developing erroneous notions upon the subject of ventilation than **the employment of unscientific** terms in connection therewith, such as *Automatic Ventilation*. Ventilation, by whatever means it is induced, is the result of certain agencies or forces, either *natural* or *mechani-*

cal, but never *automatic*. *Natural* means may be employed, or *mechanical* means, to secure ventilation, but with neither must it for a moment be supposed that the action is *automatic*. This question of terms, which is of the utmost importance to a proper understanding of the subject, is now simply mentioned as a precaution before defining what should be understood by the term "Ventilation".

Ventilation is far more than providing for change of air. Most people,-if asked what is implied by the term Ventilation, would answer—Change of air. This is true so far as it goes, but because it is only a portion of the truth, the answer is, to say the least, misleading. This can be easily proved by throwing open the windows of a crowded and overheated room on a cold and windy night. Change of air may quickly be procured, but certainly not the efficient ventilation of the apartment, because discomfort to the occupants would certainly result. In occupied rooms, something more than mere change of air is necessary to secure good ventilation, for, although in the abstract ventilation is one and the same, it may be qualified in degree, and therefore it is ever necessary to enlarge on several elementary facts, which must be clearly borne in mind and taken into account whenever an apartment has to be ventilated, or when means have to be devised whereby a whole building, consisting of several apartments of varying size and employed for various purposes under varying conditions, has to be ventilated; and in addition to these elementary facts, there are influences at work of a more subtle nature, some of them as yet but imperfectly understood, which nevertheless should be examined, so that they may be encouraged or counteracted as required.

It is more particularly with respect to **the physical properties of the atmosphere**, and its effects upon human existence, that our subject is concerned, and consequently it must be clearly grasped that, apart from exceptional or accidental circumstances,—

1. Life[1] is only sustained when atmospheric air can be freely breathed.
2. To sustain healthy life, air must be pure and uncontaminated.
3. **Vitality will be impaired**, either **temporarily** or permanently, if air be **breathed which is contaminated or is below the normal** state of purity, in proportion **to the time during which it has been breathed and the** degree of its impurity.
4. Air **is not necessarily pure even in the open** and apart from human agencies.

---

[1] Human life is here particularly referred to, but the statement applies to most living creatures in the higher scale of existence; consequently much that applies to the ventilation of buildings erected for human habitation, will also apply to stables, farm-buildings, kennels, aviaries, &c.

5. It is, however, more frequently contaminated by human agencies.

6. When air is allowed freedom to circulate, there are many processes at work by which it is maintained at almost uniform composition, and by which it is purified after contamination.

7. When inclosed within a building, it may be quickly contaminated from a variety of causes.

These are the principal reasons why ventilation is necessary, and why it should be intelligently induced; they will therefore be enlarged upon:—

1. **The action of respiration**, which continues so long as life is sustained, alternately supplies air to the lungs and expels it therefrom. The lungs, being of thin membranes formed into innumerable small cells or vesicles, which by muscular action of the body are expanded, suck in air by the bronchial tubes and windpipe, through the nose and mouth, and then, when the muscular action is relaxed, they contract and expel the air, which, however, has in the meantime been brought into very intimate contact with the blood circulating throughout the large extent of lung tissue, and has undergone a considerable change in composition and temperature, a change by which air is undoubtedly contaminated, and made less capable of sustaining a healthy existence, particularly when confined within buildings.

2. Without examining closely **the composition of atmospheric air**, it may be accepted that it is a mechanical mixture of gases—viz. oxygen and nitrogen—approximately in the proportions of one volume of oxygen to four of nitrogen, together with a varying percentage of carbonic acid; in addition to which ozone and, more recently, argon have been found in it. Vapour of water also, in varying quantities—principally regulated by the temperature of the air and its proximity to moisture—is always present. What may be the exact composition of air most suitable for sustaining healthy life we have no means at the present time of ascertaining, and the probability is that different constitutions demand for their healthy development different states of the atmosphere, and that these demands vary at different stages in life.

All investigators agree that in the composition of atmospheric air the most essential and active agent is *oxygen*; but alone, **that gas would** be too powerful in its action upon the blood, and would quickly consume life.

Nitrogen acts as a diluent, and many consider that it is limited to this office, but there are reasons for assuming that nitrogen has active although comparatively obscure functions to perform in connection with the support of the human frame.

Carbonic acid (a chemical composition formed by the combination of two parts of oxygen with one part of carbon) exists in the atmosphere in a gaseous state, and, where it does not exceed $\frac{1}{1000}$ part in volume, is not regarded as an impurity; but air known to be impure has almost invariably been found to contain larger quantities of this gas. Consequently its relative proportion has been considered a test of atmospheric purity. The most recent investigations, however, tend to show that, within buildings, it must only be taken as an index, and not that this gas itself is the actual contamination.

The definite functions of *other gases* which have been found in the atmosphere are so far unascertained as regards the support of healthy life; it may therefore be generally accepted that air containing normal proportions of oxygen and nitrogen, with a small percentage of carbonic acid, as above stated, together with a varying amount of vapour of water proportionate to the temperature, is capable of fulfilling all its functions in connection with the maintenance of life, and in that state may be considered pure and wholesome to breathe. This may be regarded as the *neutral* side of the question in respect to ventilation.

3. **Contamination of atmospheric air** by the presence of other products is the *positive* side of the question, and will claim more particular attention hereafter. Evidence is constantly forthcoming to prove that, if air be greatly contaminated, death results from breathing it, and that, even when the impurities in the air are not sufficient to cause death quickly, they may in the course of time so impair the respiratory organs, or lower the general tone of the body, that illness or untimely death results.

4. **Contamination of the outer air** may be the result of natural processes quite apart from human interference, not only in countries universally recognized as being unhealthy, but also in lesser areas even in our own country, so that mere change of one portion of such air within a building for another portion from the outside similarly contaminated will not secure good ventilation. In fact, there are districts in which at times it is more healthy to live within doors than to remain in the open.

5. The outer atmosphere is more liable to **contamination by human agencies**, which may affect whole districts, or be simply produced in or around a single dwelling or factory. In either case, it is unreasonable to expect that good ventilation can be secured by simply changing the air within a building for that from without, when the latter is contaminated.

6. Although atmospheric air is, even in the open, frequently contaminated by the presence of other gases, by vapours, and even by solid substances, nature has fortunately provided many **processes of purification**, both simple and com-

plex, by means of which the air in most localities, after contamination, again becomes pure and healthy to breathe.

Movement is perhaps the most necessary for bringing about this desirable end, for where there is stagnation some impurities increase, and often become dangerous to health. *Sunlight* also plays a very active part. Most *vegetable growth* assists by absorbing the excess of carbonic acid, and even *the mineral kingdom* exerts some influence. *Water*, in the raging storm, the soft rain and gentle dew, in many a tiny stream and brooklet, in the slowly gliding as well as in the rushing river, in the placid lake and mighty ocean, is almost everywhere assisting in maintaining the atmosphere in an uniform and healthy state. *Heat and cold* exert a powerful influence in the same direction.

Yet it must be remembered that, wherever impure matter or source of contamination exists, these purifying agencies may be neutralized, and some may even be made to assist in developing and spreading the impurities. In fact, so numerous and so complex are the agencies and processes which affect the atmosphere, either favourably or adversely as regards human existence, that the whole realm of natural science might be brought to bear upon the subject ere a complete knowledge could be gained respecting all the requirements necessary to secure the best ventilation.

7. When natural agencies are excluded from exercising their purifying effects upon **atmospheric air within buildings**, it becomes more and more contaminated, and impurities quickly increase by respiration and exhalations from living creatures, by combustion, by putrefaction of animal and vegetable matter, by fermentation, by volatilization, by particles of innumerable substances in the form of dust, by the presence of living organisms resulting from disease, and perhaps in other ways as yet undetected. The purification of the air in buildings is best secured by efficient ventilation, but where infectious disease has made its appearance in the building, antiseptic and more decided methods must be employed.

The foregoing statements as to **the complex nature of the subject** are intended to show that, if means for securing ventilation are to be employed beneficially, they must be applied with knowledge and be regulated with judgment; and also, that the ever-varying conditions and states of the outer atmosphere preclude the possibility of devising any appliance or apparatus by which constant and efficient ventilation can be automatically procured.

Advance will only be made by those who realize the necessity of pure air as a means of securing the health of individuals and communities, and who also acknowledge the intricacies of the subject, and insist upon the fact that constant individual attention is required (even when reasonable means are provided), if good ventilation is to be continuously secured. Those who have never studied the subject are ever the most ready to affirm that it is a very simple matter, and to condemn architects and builders for not providing everlasting means whereby, under the most varying conditions, a supply of fresh air, exactly to their liking, can be conveyed to them without trouble or further expense, regardless of the facts that the state of the atmosphere is constantly changing, that apartments are variously occupied, and that individuals are so constituted or affected by habit, that an atmosphere in which some can comfortably exist, is offensive or discomforting to others. This unreasonableness on the one hand, and on the other the persistency with which many disregard the necessary provision to secure ventilation, or even decline to provide for it because of the difficulties it presents, are responsible for retarding true progress in the matter.

To secure ventilation is much more than a personal undertaking, because it is useless to expect that good ventilation can be obtained unless the outer atmosphere is in a state of purity; and although many things have been accomplished during the present century, in a communal manner, to make it more possible than it was to secure a sufficiently pure atmosphere in and around our dwellings, the fact still remains that, where communities are thickly congregated in cities and towns, constant vigilance is required on the part of Local Authorities to prevent the carelessness of some and the apathy of others jeopardizing the general health by uncleanliness, or by permitting the accumulation or discharge of impurities resulting from domestic requirements or manufacturing processes.

The State and Local Authorities have in recent years and in many ways attempted to prevent contamination of the outer atmosphere, by regulating the discharge of smoke and noxious gases, by preventing the pollution of streams, by drainage, by the removal of house-refuse, by the cleansing of streets, by regulating the keeping of animals in and about dwellings, and by the notification and isolation of infectious diseases; but there is still much that might be done, not only in the courts and alleys of cities and towns, but also in better-favoured localities and country places, before the outer atmosphere can at all times be considered reasonably pure and wholesome; and even when all has been done that can be expected of the State and of Local Authorities, it is still incumbent upon every individual, by habits of personal and domestic cleanliness, and by a knowledge and daily use of the means of ventilation, to safeguard both his own health and that of those residing around him. Regarded thus, ventilation is a collective as well as a personal matter.

If every individual appreciated the value to health which results from constantly breathing pure air, no cost would be spared in securing it; yet because the effects of living in an impure atmosphere are only at times, and by a few highly-sensitive people, made quickly known, the majority are callous, and begrudge the outlay and trouble which are necessary to supply them with wholesome air. They hourly take into their systems an insidious poison which, although it may be slow in its effects, will certainly cause suffering and probably untimely death. Even those who would be disgusted at the idea of partaking of food not perfectly pure and fresh, or of drinking fluids which have become contaminated, and would on no account eat or drink what they knew had been near the lips of anyone else, will nevertheless be content to live in a close and unwholesome atmosphere, and freely breathe the air which has recently been respired by others, regardless of the consequences; and yet, because of the necessity of constantly supplying the lungs with air to sustain life, the large volumes of impure air inhaled will more surely undermine the constitution than the occasional partaking of food or drink, which to their sense of taste, sight, or smell, may appear to be contaminated.

Let us now consider **what other conditions are requisite for the comfort of householders** for whom ventilation is demanded. Buildings are erected principally as a protection from the ever-varying conditions of the atmosphere; this implies that comfortable ventilation is not at all times to be secured in the open, even presuming that the air is pure. If in the open we cannot secure comfort, and consequently build houses because at times movement of the atmosphere is too rapid, at other times too stagnant, now too hot and anon too cold, sometimes too dry and often too wet, it is evident that by enclosing a portion of the air within walls and roofs a risk is run of causing stagnation; free circulation at all events is impeded, and most of the influences, which outside are at work upon its purification and maintenance in an uniform state as regards composition, are shut out; hence the necessity for special *means* being provided whereby the air within buildings may be constantly changing, and maintained in a condition suited to the comfort and capable of sustaining the health of the inmates.

CHAPTER II.

NATURAL FORCES MAKING FOR VENTILATION.

To change the air of a building two things are primarily essential: (1) an air-inlet, and (2) an air-outlet.

This may appear so obvious a requirement that no more need be said thereon, but although it may in the abstract be generally known, it seems to be only partially recognized and is frequently ignored in practice.

In opening a window to obtain more air, as it is termed, in an apartment, is it always recognized how that which was previously there finds exit? Air entering by the window is felt, or is known to be coming in by the movement of objects in its vicinity, but inasmuch as the room was not previously void of air, and as other air could not enter without displacing that which was previously there, some means of escape there must be. If, then, by opening a window appreciable change of air in an apartment takes place, several considerations arise:—

1. A contrivance, i.e. the window, made to open, had been provided, partly at least (we may presume) for the purpose of ventilation.
2. A personal act was performed, viz. the opening of the window.
3. There must have been some other outlet or inlet communicating directly or eventually with the outer air, although neither may be well-defined nor clearly visible.
4. According to the size and position of the openings, both for inlet and outlet, a given change of air in the apartment would take place under similar conditions.
5. If we presume that the apartment was previously occupied, there must have been some change of air going on before the window was opened.
6. If this incoming air entered directly from the outside, where was the inlet, or more probably inlets? Were they in proximity to anything by which the air might be contaminated, such as drain-gullies or accumulations of refuse, the household midden or dust-hole, or rain-water-pipes with open joints and improperly connected with the soil-drains?
7. If the air had previously traversed some portion of the building, where did it first enter, what direction did it take, and what may it have passed on the way? If pure on entering the building, could it have been contaminated on its way to the apartment?

8. Was the change effected by a force acting from without, or by a force acting from within? If from without, was it a *propelling* force, which drove the air in and so displaced that which preceded it; or was it a *suctional* force, which extracted that from within and gave place for other air to enter? If from within, what force could have been brought into action which would cause either a forcing-out or a sucking-in of air?
9. Having presumed that the opening of the window was necessitated for the greater comfort of the occupants of the room, it follows that, previous thereto, change of air was insufficient, or that the source from whence it was drawn was not a pure one.
10. Did the opening of the window lessen or completely stop the entrance of air from other sources? Was the air entering at the open window uncontaminated? What was its temperature and rate of movement?
11. Could the occupants continue in the room with comfort while the window was open, or would other discomforts, such as draughts or excessive cold, compel them to shut it again?
12. If after an interval the window was again closed, and the occupants experienced greater comfort for a time than they did before it was opened, what deductions must be drawn?
13. Might not the regulation of the window-opening from the first have secured comfort to the occupants? If not, could other openings in any other part of the room have provided the requisite change of air, without discomfort, at the particular time and under similar circumstances?
14. If so, would a change of conditions, or a different state of the outer atmosphere, necessitate further regulation, or the closing of such openings?
15. In order to secure comfort for the occupants, was there a fire burning, or some other means of raising the temperature of the room?
16. What was the cubical capacity of the room?
17. What were the number of the occupants, and their occupation?

Although the foregoing list by no means exhausts the questions which may arise even under such ordinary circumstances as have been supposed, it is sufficient to indicate the following general principles:—

1. That, after ascertaining the possibility of procuring reasonably pure air from the outside, constant care will have to be exercised in order that it may not be contaminated either just before or while entering the building, or after having entered.

2. That openings must be provided for its entrance and for its exit, with means for regulating either one or both.

3. That the regulation of such openings requires frequent and intelligent attention.

4. That when change of air is brought about within a room it is by means of forces which, if natural, are constantly varying, and therefore when such natural forces are employed, the rate at which change of air takes place must necessarily vary.

A mechanical power may be employed, which would be more regular in action than the natural power of wind, or than the force exercised by the air within and that without being of varying density, caused by difference of temperature, but up to the present time only a few ordinary dwellings have been ventilated by mechanical means, and there can be little doubt that only in exceptional cases can mechanical means be employed for economically ventilating isolated houses of moderate dimensions. For public halls, churches, libraries, hospitals, schools, factories, and other buildings erected for public purposes, particularly those in which many people congregate, ventilation by mechanical means alone can be depended upon to give constant and satisfactory results. Its use may also with advantage in time be extended to dwellings built compactly together or in flats. The principles of mechanical ventilation, however, require the utmost knowledge and care, and their application should only be attempted by those who have made a thorough study of the subject.

As the employment of mechanical means for ventilating single dwellings is, and is likely to be, of rare occurrence in this country at least for many years to come, it is not proposed to enter further into the subject at present, and, therefore, **the forces which Nature supplies will now be considered.** It cannot be too strongly insisted that change of air within a building is invariably brought about by the exercise of a power capable of being ascertained. This power can be traced back to the effects of *heat*, either of the sun or arising from the combustion of fuel, and that the effects of heat are made operative by the force of *gravitation*, which, apart from the employment of mechanical means, may justly be regarded as the active agent by which change of air is brought about naturally to assist ventilation. Knowledge of this fact simplifies the whole subject, and explains much that otherwise may appear complex. There are certain other natural properties of substances which exercise influence upon the atmosphere, and to some extent regulate air-movements, but they are comparatively insignificant compared with the force of gravitation.

A property of the atmosphere, which is, however, common to most gases, is that heat-rays pass through it without materially raising its temperature, and yet when air comes in contact with either solids or liquids, the temperature of which differs from that of the air, the latter will either absorb or give up heat; this property exercises a powerful influence in creating movements in the atmosphere, and consequently upon the ventilation of our houses.

When its temperature is raised, air expands, and becomes, bulk for bulk, lighter than air at a lower temperature, the result being that cold air falls by the power of gravitation, and in doing so causes the lighter air to ascend. As this process is going on at all times to a greater or less degree, throughout the whole surface of our globe, movements of air, constantly varying in force, are set up. In some districts this movement, brought about by the great heat of the sun, is so rapid as to cause violent storms, but generally the action is more regular, resulting in what are known as trade-winds; yet, as day succeeds night, and many local conditions induce variations in the earth's temperature in different places at the same time, even these comparatively regular trade-winds are variously affected, resulting in what are popularly known as changes in the weather.

To a very large extent, it is upon these most variable movements of the outer atmosphere that we must depend for securing ventilation within our homes. The forces they exercise may be employed either for propulsion or for extraction, or for both at one time. In warm weather, when it is possible to throw open doors and windows freely, the propelling force of wind is almost exclusively relied on to obtain change of air within a building, but when doors and windows are kept closed, the suctional influence of the wind, passing over the open chimney-pots and air-openings, has a considerable influence upon change of air within.

In addition to these forces from without, when houses are warmed either by open fires or any other means, additional forces are developed within, which may either assist those outside in causing change of air within, or may be antagonistic thereto. Smoky chimneys are evidence of this antagonism. The forces set up by the employment of heat within a building are caused by the interior atmosphere taking a higher temperature than that without, whereby it expands, and, being lighter than an equal volume of the air outside, is forced upwards by the colder and denser air descending.

Just as the effects of variation in temperature outside produce movements of air, which may be simply local or of vast extent, in the same way but in a lesser degree will varying temperatures within a house or even an apartment cause local or general movements in the atmosphere within. Such local movements must not be confounded with actual change of air, because local movement or circulation within a building is not sufficient to bring about that change of air which exercises the purifying effect required for the purposes of good ventilation.

The British home is the stronghold of **the open fireplace**, and this, on account of its cheeriness and its assistance in ventilating the house, is likely for long to find an honoured place. A valuable amount of ventilation is secured, not alone by the presence of the fire, but to a large extent because of the necessity for a smoke-flue or opening to the outer air of adequate size. Many people overlook this fact, and when no fire is placed in the grate, most unreasonably close down the register, stuff something up the flue, or place a screen over the grate-opening, and in so doing prevent that free change of air in the apartment, which almost invariably goes on so long as the flue remains open. Such practices cannot be too rigorously condemned, for whether there be a fire in the grate or not, the flue from every room should be regarded as a most important outlet for air, the exit of which will alone permit other air to enter.

Because of the more general tendency of **up-draught in flues**, there is a belief that they exercise a suctional influence, but the fact is that, wherever such suctional effect is noticeable, there is an actual force to cause it, the flues being simply channels by which the resulting movement of air is made apparent.

In like manner it has been thought that with flues of varying height, connected at the base, the longer one will exercise **a syphonic influence**. When a downward current is observed in the shorter flue, careful examination will demonstrate that it is the result of some disturbing cause in the atmosphere itself, or of some force applied, without which the flues would of themselves be quite incapable of influencing the movement. It would be as reasonable to expect water to circulate within a syphon plunged into a vessel of water. The material form of the syphon or of the flues cannot, of itself, exert any influence which would cause a movement therein. Nevertheless it is a fact in practice, that what appears to be a syphonic action does frequently take place where there are flues of varying lengths in air-connection at the base, but it is the result either of variation in temperature, or the suctional or propelling force of wind, or of mechanical power, and it would be quite easy to apply any one of such powers to cause the air to ascend the shorter flue and descend the longer one, although under ordinary circumstances the reverse will take place.

Ordinary dwellings in this country are more frequently warmer than the outer atmosphere, and open fires are commonly employed during the colder months of the year; consequently, unless each room receives a separate and adequate supply of air direct from the exterior, there is **risk of down-draught in some flues**, particularly if others are more lofty. It is therefore advisable to make as little variation as practicable in the height of the flues in a single dwelling, although, even if all the flues are carried to the same height, they will draw on one another if there is variation of temperature in the several flues and air-communication at the base, without separate air-supply to each apartment. Frequently this is the cause of annoyance resulting from "smoky chimneys", i.e. of flues with a down-draught tendency, which probably could be cured by admitting a supply of air from the exterior to each room separately.

The section of a house, shown in fig. 561, will illustrate this. A fire lighted in either of the rooms would have a tendency to draw air by way of the staircase down one of the other flues, provided that the windows, doors, and other openings to the exterior are closed. This other flue may even suck down smoke emitted from the flue at the base of which there is a fire. On attempting to light a fire in the other room, it will be found that smoke will not freely ascend the flue. Close the doors of both rooms and partly open the windows, and if smokiness within the house resulted simply from the suction of one flue upon the other, the fires will both burn briskly and without smokiness.

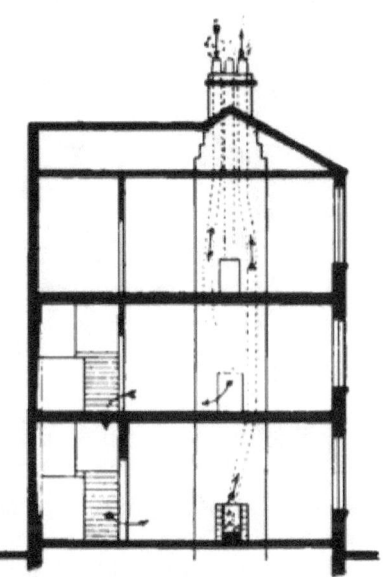

Fig. 561.—Section of House, showing Danger of Down-draughts in Flues when Air-Inlets are not provided.

A central hall provided with an extract-ventilator at the top, as in fig. 562, will, either from the air thereof becoming warmer than in some of the rooms around, or by wind acting upon the outlet, have a tendency to draw air from the rooms, so that, unless there are other inlets, air will be sucked down the flues and result in a sooty smell about the premises.

These two examples alike indicate **the necessity for air-inlets**, relatively proportioned to the outlets, to each separate compartment throughout the building.

As change of air is essential to ventilation, there must be movement of air.

When this movement is too rapid, or is ill-directed, uncomfortable draughts are the result. Even warm air moving quickly will cause discomfort, and may be more pernicious than colder currents, because, unless such warm air is in a proper hygrometric condition, it will cause excessive evaporation from the body, and by evaporation a lowering of temperature takes place. To obviate draughts and yet give the necessary change of air, substituting that which is

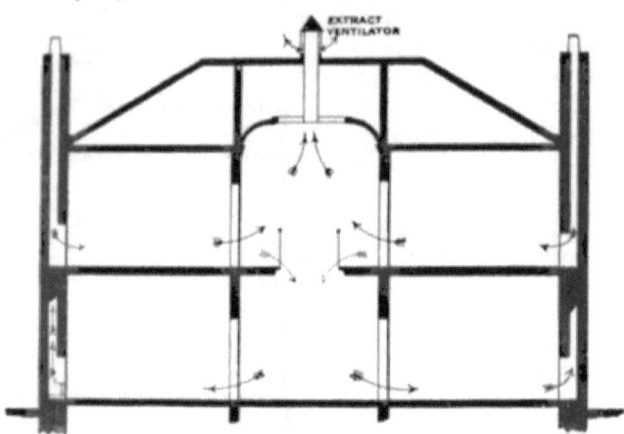

Fig. 562. Section of House with Extract-ventilator, showing Danger of Down-draughts in Floors when Air-inlets are not provided.

fresh and wholesome for that which has become vitiated, is essentially good ventilation.

The sizes, nature, and positions of inlets and outlets must be carefully considered in all schemes for ventilation. Supposing that a force be employed for propelling air into an apartment which measures 20 feet by 15 feet by 10 feet high (equal to a cubic capacity of 3000 feet), so as to change the air six times in an hour, a volume of air equal to 18,000 cubic feet per hour would be required; this might be introduced through an aperture one foot square, if travelling at the rate of 300 feet per minute = 5 feet per second. The principal outlet is supposed to be the fireplace flue. The question will arise, where should the inlet be situated? If placed on the opposite side of the room, as in fig. 563, the incoming air will take the most direct course to the outlet, and between the two there will be considerable draught, and far more rapid change than in other parts of the room. If, however, the opening be trumpet-mouthed inwards, there will be better diffusion; but this will be still more improved if such an inlet is placed on the same side of the room as the outlet, as in fig. 564, and the risk of setting up unpleasant draughts will be reduced to a minimum, unless the incoming air is at a very low temperature.

Special air-inlets, other than windows and doors, should be situate at about two-thirds the height of the room, and shaped so as to give the incoming air a

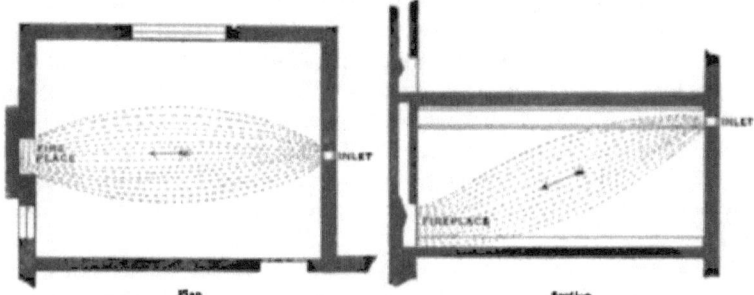

Fig. 563.—Plan and Section showing Air-current in Room, with Square Inlet opposite the Outlet.

slight upward tendency; then, if they are placed on the same side of the room as the outlet—viz., the fireplace, which is near to the floor-level—the incoming air will distribute itself about the upper portion of the room. With a fire in the grate the surfaces of walls, ceiling, furniture &c., will be warmer than the

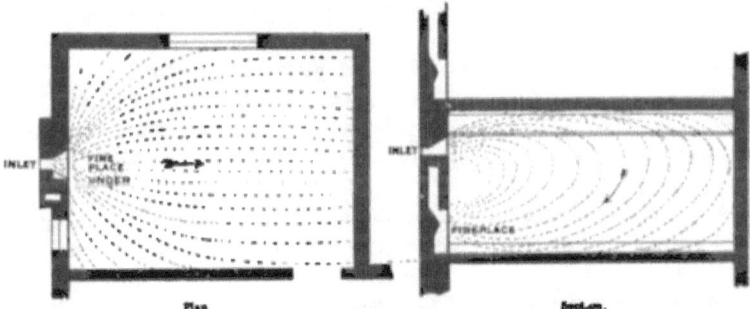

Fig. 564.—Plan and Section showing Air-current in Room, with Trumpet-mouthed Inlet on the same side as the Outlet.

incoming air, and this, by contact with such surfaces, and by mixing with some of the warmer air of the apartment, will acquire a higher temperature than when it entered; it will have to travel a considerable distance within the room, whereby its velocity will be reduced, and if the inflow is continuous, and there are no other disturbing influences, it will advance at a low rate of speed towards the outlet, through the lower portion of the room where occupants are most likely to be, and, as regards its temperature, in a condition which will tend to their comfort. As it approaches the fire, the suctional influence exerted by the rarefaction of the air passing up the flue will gradually increase its velocity until it reaches the mouth of the flue, up which it will rush to the open.

It unfortunately happens that many eminent writers on ventilation, without carefully distinguishing between the kind of ventilation required, advocate the placing of air-inlets and outlets on opposite sides of rooms. It is true that to get good through ventilation, openings, such as windows and doors, should be situate on opposite sides of the same apartment, but for comfortable ventilation, particularly in cold weather, with a fireplace flue for the outlet, the inlet should be on the same side of the room. This also holds good whenever mechanical means are provided for securing ventilation, whether by *extraction* or *propulsion*.

If, however, neither open fires nor mechanical means are employed, and movements of the outer atmosphere are depended upon to secure change of air within a room, and the inlet and outlet openings are in outer walls, then undoubtedly they should be on opposite sides if practicable; if not, then on adjoining sides, because it would be difficult to form both inlet and outlet through a single wall in such a manner as to induce a constant current in one direction. The pressure of wind would generally be practically equal on the two openings at the same time, or at times it might vary, first acting on one, then on the other, so that at best an intermittent flow of air would be induced.

Experiment has demonstrated that air at normal temperature, moving with a velocity exceeding 5 feet per second, is generally felt as a draught, at 4 feet most people would consider it uncomfortable, at 3 feet some would complain, but at 2 feet few would notice it, and with a velocity of 1 foot per second it would not be perceptible to anyone.

If air enters the room shown in fig. 564 at the rate of 300 feet per minute— *i.e.* 5 feet per second,—its **velocity** may be reduced as it enters the apartment, by belling out the inlet-opening; the air will then quickly be diffused in the upper portion of the room, and, if there be only one exit-opening, no larger than the inlet, the velocity of the air in approaching it will increase to its original rate, but by enlarging the embouchure to (say) 4 square feet, the rate of air-travel will at that point not exceed $\frac{300}{4} = 75$ feet per minute, or $1\frac{1}{4}$ feet per second, and will be considerably less throughout the habitable parts of the room.

This is presuming an air-tight apartment, but as in practice such is scarcely attainable, it is usual, when propulsion is employed for securing ventilation, to provide an outlet somewhat smaller than the inlet to allow for leakages, which, if **small and** well distributed, may tend to a more thorough distribution of the **incoming air,** and reduce the velocity at the main outlet.

By constantly keeping in mind these simple facts as to the relative positions **and forms of inlets and** outlets, discomfort from draughts may be, to a great **extent, avoided, although,** without mechanical power, it may not be possible to regulate the changes of air with absolute uniformity. The great aim of ventilation is, however, not to secure mathematical accuracy in the quantity **of** air supplied, but to obtain an adequate (even though variable) **change of air without inducing objectionable draughts in the apartments ventilated.** Whether this be done by mechanical or natural means, or by a propulsive or suctional force, matters little so long as the current continues in one direction; and it may be laid down as a general rule, applicable in almost every case, that the air-inlet and the air-outlet should be on the same side of the room, and that both should be belled out towards the room, the air-inlet being best placed at about two-thirds of the height of the room, and arranged so as to give the air a slight upward tendency, while the outlet should be at or near the level of the floor.

The problems of warming and ventilation are so interdependent, that the one subject cannot be fully considered without reference to the other; thus, ventilation is simplified when some form of warming apparatus is adopted in addition to open fires. It then becomes possible, by means of suitable radiators, to warm the air entering the room and thus to prevent cold draughts, which are often dangerous to health. More will be said of this in a subsequent chapter.

The following simple experiments will illustrate some of the principles laid down in this chapter.

1. To show that change of air is principally brought about by variations in temperature, take a large glass vessel, fill it with pure air, and insert a few light particles, such as down; close it so that the air within cannot be changed, and by alternately warming different parts of the vessel, the air within may be made to circulate, the circulation being made apparent by the movements of the particles of down.

2. If a living animal or lighted taper be inserted, and the vessel again closed, it is only a question of time when the one will die and the other be extinguished, by exhaustion or contamination of the air within; a similar result may also take place even in a vessel with a single opening at the top, although a longer time will be required; but with an opening also in the bottom of the vessel, the taper would freely burn and the animal could live, thereby proving that actual change of air takes place, resulting principally, if not entirely, from the heat evolved either from the lighted taper or from the living animal within.

3. A similar glass vessel may also be employed to illustrate the varying effect of heat and cold upon the hygrostatic condition of air. Let warm air be admitted after it has had free access to moisture, then if the vessel be closed and placed in a colder position, it will become bedewed inside, or if only a portion of the vessel be considerably cooled down, dew will be formed within upon that colder part.

4. Further effects of condensation will be seen if the air within the vessel be contaminated with volatile organic matter while it is warm—say that a taper has been allowed to burn therein for a short time;—then if the vessel be closed and cooled down, a film will be deposited within consisting of the products of combustion and volatilization of substances composing the taper.

5. It is also useful to note that, if in either of cases 3 and 4 the vessel be again heated, the moisture or the film of animal matter will once more be volatilized and become invisible.

CHAPTER III.

THE PLANNING AND CONSTRUCTION OF HOUSES WITH REGARD TO THEIR VENTILATION.

The aspect and situation of a house may have a decided effect for better or worse upon its ventilation. If overshadowed by more lofty buildings, elevated ground or trees, free circulation of air around may be impeded, or the revivifying rays of the sun may be shut out. The position of a house with regard to others even of a similar class may likewise affect ventilation.

Wherever it is practicable, **windows** should be provided on more than one side of a room, it being better for the cheerfulness of the apartment, as well as for the purpose of ventilation, to distribute them than to have all the window-space massed together. The literal meaning of the word window is "an opening for the wind", and although windows are now more generally regarded as means for admitting light to the interior of buildings, they ought nevertheless to be constructed so as to be capable of being easily opened, and should still be considered the principal means for securing change of air in an apartment. When there are more than one, on different sides, it is possible to utilize outer movement of the air to cause change within, quite independently of other openings, and although, when the room is occupied, it may not at all times be comfortable to permit of such rapid movement of air within as may frequently be induced by the opening of windows, there are times during which rooms are unoccupied, when it may be allowed with great advantage. Hence the advisability of carefully considering, during the planning of a house, the relative positions of windows and doors with a view to securing a thorough change of air throughout the building or in individual rooms at convenient times.

Bay-windows are particularly adapted to securing comfortable ventilation in hot weather, because, by a considerate adjustment of curtains and blinds, it is generally possible to have the window open on one side at least of the bay, and yet to shut out the glare and heat on the other sides.

Detached houses should be situated in reference to their surroundings, so as to facilitate free circulation of air around, and the doors and windows should be so placed that advantage may be taken of the best aspect for the several apartments, and that, when desired, windows may be opened freely to catch the most favourable breezes. Through ventilation ought then to be secured, for, when fresh air surrounds a building, it is possible, with a little care in the opening of windows, to obtain a movement of air through the building at almost any time. If the wind be strong outside, a moderate opening of windows will suffice, and, even when the air outside is sultry, by freely opening them on opposite sides refreshing movement may generally be secured within, brought about by the varying temperature on one side of the building and the other.

Houses in rows or terraces are not so readily ventilated at all times as detached houses, because air has not free access all round them. Much, however, may be done in their planning so as to secure a reasonable amount of through ventilation, because the difference in temperature between one side of the houses and the other at the same time will, at most hours of the day, cause a movement of air through them if windows be judiciously opened.

Houses in courts and alleys are more difficult to ventilate properly; consequently their erection should not be permitted. Fortunately they are now prohibited in all well-regulated localities.

Back-to-back houses are even more objectionable, and should be condemned.

Lofty houses, especially if built close together, are not only difficult to ventilate, but, in addition, they overshadow so much ground-surface or other buildings in their vicinity, that purity of air around, and consequently good ventilation within, cannot readily be maintained. In other words, overcrowding of a given surface of ground must be avoided if efficient ventilation of dwellings is to be secured.

If buildings are erected of impervious materials, change of air within can only be adequately effected by providing definite entrances and exits. This of itself might be an advantage if these entrances and exits could be so arranged as to cause the incoming air to disseminate through every portion of the building, because then we could the more readily regulate the quantity of air which should enter and leave the building as well as determine whence it came. The difficulties, however, of securing suitable appliances, and that frequent personal attention which would be necessitated by the ever-changing conditions of the external atmosphere, lead me to believe that such buildings are not likely to be the best ventilated. Moreover, there are positive disadvantages in the use of impervious materials, because, as a rule, such materials do not retain heat; they may become quickly heated and as quickly cool; consequently variations in the temperature outside soon affect the condition of the air within. There are, however, other impervious materials which do not readily absorb heat, but reflect it, and are therefore cold to the touch. In either of these cases, there is always a tendency, when the outer atmosphere is colder than that within, for moisture and volatile substances, as well as organic matter, to be condensed upon the impervious materials. The author has frequently noticed that in rooms the walls of which are covered with varnished paper, and where the floors are impervious, it is most difficult to obtain ventilation without draughts.

Some may suppose that the pores of pervious walls would become contaminated by the passage of air carrying with it impurities. This might occur in localities where the atmosphere was very impure, but then it would be equally injurious to admit the air by a window or other opening. Under ordinary conditions, impurities in the air would be deposited near the surface, and would be subject to the generally purifying effects of the outer atmosphere.

Houses built of excessively porous materials, on the other hand, may be much more unhealthy, because, in rainy weather, the tendency is for them to absorb much wet, which, if it does not actually penetrate to the interior and do definite damage to the walls and decorations, produces a chilling effect within, principally brought about by evaporation, which always results in a lowering of temperature and causes condensation within, as previously described.

My opinion is that both extremes are wrong, and that, if more attention were paid to the selection of suitable materials for the walls of our homes, there would be little call for special ventilating appliances[1],—other than windows and fireplace-flues,—particularly in localities where cleanliness reigns within and without, where no overcrowding takes place, and where care is exercised in opening windows freely at suitable times. On the other hand, it is unreasonable to expect that efficient ventilation can be secured in badly-constructed dwellings,

[1] One of the chief advantages of "ventilating appliances" is that they can be easily regulated.—ED.

built with unsuitable materials, or where there is neither cleanliness within nor without, or where overcrowding is permitted.

The construction of the internal partitions both horizontal and vertical, and the condition in which they are left in those portions generally out of sight, have much to do with movements of air within, which do not necessarily result in efficient ventilation. In fact, it may as a rule be considered that such movements of air from one apartment to another are objectionable, and are often most harmful, if there is air-contamination within any part of the building, which can be communicated to other parts.

Let us now make an imaginary examination of a house of ordinary type, erected (say) a few years ago, before sanitary science had received so much attention as of late. This I suggest because I have learnt by experience the advantage of examining closely the causes of defective ventilation, an intimate knowledge of which will enable one at times to remedy the evil with comparatively slight trouble and at small outlay.

We will select a moderately cold day, and enter an apartment in which a fire is burning briskly. Close all windows, doors, and inlet-ventilators, and hold a lighted taper to the key-hole in the door, to the cracks around door and windows and under the skirting board, and to the cracks in the floor-boarding, and probably the inlet of air will be detected, more or less, at each.

Stop all noticeable crevices and cracks, and a further search will detect others by which air can enter, and even if they can be closed, it will still be found that air must be entering to replace that which is continually ascending the flue; although probably, if the closing of inlets has been thorough, the fire will not be burning so briskly, nor will the velocity of the up-current in the flue be so rapid, but some change of air must still be taking place in the room.

Many people hope to lessen draughts by closing all cracks and crevices, but by doing so they are more likely to create others more cutting, unless suitable provision be made for the admittance of an ample supply of air in a suitable position, so that it may at once be equally diffused throughout the apartment. This can readily be illustrated in a room with a fire burning and all inlets from the outer air closed. Slightly open the door, and probably anyone near the opening will experience an unpleasant draught, but fully open the door, and unless there is a direct blow of cold air from the outside, no perceptible draught will be felt, although a larger volume of air may be passing through the room than when the door was only partially open.

Let us continue our examination of the room, supposing it to be on the ground-floor, with no basement under. Inspect the space under the floor-boards.

Is it clean and freely supplied with fresh air? If not, that drawn into the room will be more or less contaminated. Air-circulation there must be under wood floors, or dry-rot will set in; care, however, must be exercised so that air-gratings for providing ventilation under ground-floor rooms be placed so as to avoid drawing air from impure sources, and householders should be ever on the watch to prevent the accumulation of filth in the neighbourhood of their houses, particularly near air-inlets. The continuous ventilating anti-damp course[1] has much to recommend it as an anti-damp course, and as a means for providing an air-current under all floors, but, on account of the modern requirement that all wastes discharge visibly into gully-traps, and the chance of accumulations of filth being permitted around some portion of a building, I am of opinion the better plan is to make the anti-damp course solid, and to select positions for the air-grates where contamination of the air is least likely to take place.

In examining a room on an upper floor, consider first the construction of such floor. Remove some of the floor-boarding, which perhaps is not tongued and grooved, so that each joint is open, and if not very wide, may in part have become stopped up with accumulated filth. The space between the floor and ceiling of the room below is probably open. If meals have been partaken of in the room, crumbs and organic particles resulting from the habitation of human beings, animals, &c., may be detected among the dirt and dust which will have accumulated. Washing the floor will moisten these particles, which will then putrefy and throw off deleterious gases, to be drawn into the apartment through the cracks and crevices in and around the floor. Moreover, the plaster ceiling is porous and often cracked, so that, when the joints of the floor-boarding are also open, there is actual air-communication between the two rooms, and even if direct currents from one to the other are not set up, the property of diffusion of gases will certainly come into play and cause the air of the two rooms to mingle.

The lower rooms of a dwelling are generally employed as sitting-rooms, those above as bedrooms; consider what may, under such conditions, take place during the winter months. Several people may occupy the ground-floor rooms, and a considerable quantity of gas may be employed for lighting purposes. Windows, doors, and ventilators may be closed, and even under moderately favourable conditions of ventilation, it will be found that the air of the top portion of the room will become so highly vitiated that it cannot be breathed for long with comfort. The tendency of this foul air will be to make its way through ceiling and floor to the rooms above, to which, after a time, some of the occupants of

[1] Figs. 29 and 30, vol. i. p. 85.

the sitting-room will retire to sleep, and will have again to breathe the same air, which may already have passed through their lungs and in other ways have become vitiated.

Now notice where the joists pass through the inner walls or partitions, as there may be openings through to the adjoining rooms; trace them along, and if there is a water-closet on the same floor, you may find a clear air-way through to it between the joists. If there is any accumulation of filth near the closet-basin, or if any of the joints in connection with the closet and branch soil-pipe leak, the foul air may be drawn along the floors to other rooms, especially when fires are burning in them.

The space usually inclosed under the roof, and not utilized for dwelling purposes, becomes an air-chamber, generally connected in a more or less direct manner with the several apartments. Too often such spaces become receptacles for all sorts of lumber, and are frequently allowed to become extremely dirty. In such cases, air drawn into the rooms therefrom can scarcely escape contamination.

The construction of the roof has a marked influence upon the temperature within a building. A mere covering of slates affords little protection except from wet, and in many cases not even from that. Then it is that the walls and ceilings become saturated with water, and cold and dampness, decay, and consequent deterioration of the air of the dwelling follow. In summer slates become heated and communicate heat to the air below, and in winter the reverse takes place, by which the ventilation of the building is constantly being affected in an adverse manner. When the slates are laid with very open joints, even an ordinary wind will blow through and cause discomfort, but in stormy weather the effect may be to rob the whole building of the greater portion of its heat, and to cause draughts which are positively unbearable.

Well-made plain tiles, which, while not being absolutely porous, should not be too much compressed in the process of manufacture, or they will partake more of the properties of slate, afford better protection against changes of temperature without; yet having regard to the efficient ventilation of a dwelling, both slates and tiles should be laid upon a non-conducting material, such as boarding, felt, pugging, or the like, in order to prevent changes of temperature outside having any rapidly-marked effect upon the air within. Care must, however, be taken to secure reasonable change of air within the roof-space, independently of the apartments of the house intended for occupation.[1]

[1] For illustrations of houses specially built for the purpose of enjoying thorough ventilation, see Plates XX., XXI., and XXII.; some account of these houses will be found in a supplementary note to this section on pp. 224-227.—ED.

CHAPTER IV.

AIR-CURRENTS AND AIR-INLETS.

Before deciding upon the position of a special air-inlet, particularly to an existing room, it is important to ascertain the direction of the air-currents in the room with a view to ascertaining that position from which the incoming air will be most equally diffused without creating draughts. This is not always easy to ascertain, because of the invisible nature of the atmosphere, and the very slight influence it exerts when its movement is slow.

One method adopted to assist in the detection of air-currents is the employment of volatile essences, the odour of which may be easily recognized, such as oil of peppermint. Their power of diffusion, however, is so rapid, and the course of air-conveyance at times so obscure, that it is difficult to ascertain with certainty, by their use, the course of air-currents. Smoke is also employed for the purpose, but the heat evolved in the production of the smoke often determines its direction upwards when otherwise the movements of the air have a horizontal tendency. If air-currents are moderately strong, a lighted taper is useful as a test by observing the deflection of the flame.

In some houses it will be observed how much more quickly and completely kitchen-smells are conveyed throughout the building than in others; this generally results from the want of independent ventilation to the several apartments, although in some instances the evil arises from the general arrangement of the plan and the relative positions of fireplaces and doors.

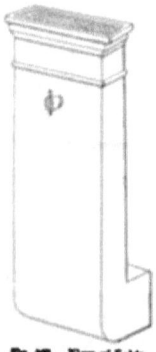

Fig. 565. — View of Tobin Air-Inlet Tube.

The provision of separate air-inlets for every room is undoubtedly a great step towards preventing the air from one room (the kitchen, for example) being drawn into any of the other rooms. The principal varieties of such air-inlets will now be described and illustrated.

Tobin tubes (fig. 565) have been largely used for air-inlets, but as usually supplied and fixed they cause so much discomfort, that frequently they are kept closed or quickly removed. One disadvantage they have which has not received sufficient attention: viz. the external gratings are, for ground-floor rooms, often placed too near to the ground-level, and at times I have found them situated close to sources of impurity, such as gully-gratings and accumulations of refuse. There is not the slightest necessity for a long vertical tube;

the direct upward tendency given to the air passing through such a tube at a high velocity causes it to rebound from the ceiling and come down like a cold shower upon the occupants of the room. Wall-papers and ceilings also become disfigured by dirt brought in with the air. The length of tube is also liable to become fouled.

Fig. 566 shows a short form of Tobin tube, with a valve A, perforated baffle-plate B, and a projecting piece C to prevent discoloration of the wall above. Canvas bags are also provided for inserting in these tubes, but unless they are regularly cleaned, they become very dirty, and prevent the entrance of air.

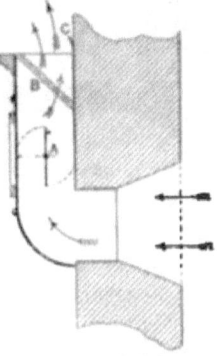

Fig. 566.—Short Tobin Tube.
A. valve; B. baffle plate; C. lip to prevent discoloration of wall above.

Simple openings through the outer walls, expanded inwards and provided with a flap or louvre-regulation, suitably placed on the *same side of the room as the fireplace*, are by far the best air-inlets which can be supplied to supplement window-openings, because they are simple and easily kept clean. So long as it is compatible with comfort to admit external air direct to the room, without *first* raising its temperature, the air may be admitted by such openings, and be equally distributed throughout the apartment in sufficient quantity.

As a rule inlet-openings are far too small, and consequently when the extract power in the outlet-flue is considerable, air enters with too great velocity and causes draughts, or air is drawn from other sources. The sectional area of an ordinary chimney-pot is about half a superficial foot, and the velocity therein will, with a fire burning, often be at the rate of 600 feet per minute, which is equal to a withdrawal of 300 cubic feet of air from the room in that space of time. In order that all the air required should be admitted by the specially-provided air-inlets, and that on entering the room the velocity may not exceed 5 feet per second, an absolutely clear opening of 144 square inches should be provided for a hole straight through the wall, but with a grated hole increasing in size from the outer to the inner face, the outer opening might be 15 inches by 12 inches, filled with an iron or terra-cotta grating, with perforated zinc of $\frac{1}{4}$-inch mesh on the inside, together with a regulating flap. The opening on the inner face of the wall might be 24 inches by 18 inches, and would be best kept clear, but might be inclosed by a wire-guard fixed with buttons, so as to be easily removed. Such openings would be little noticed if coloured to the same tone as the walls of the apartment, and they can easily be kept clean. If placed just below a picture-rail, or provided with a slight ledge above, as shown in fig. 567, marking of the walls by the incoming air will be practically avoided. For ordinarily calm weather, the inner flap should be entirely open, but when strong

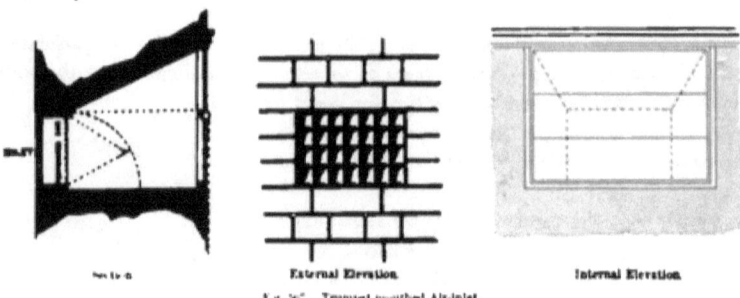

Fig. 567. Trumpet-mouthed Air-inlet.

winds are blowing against the outer face of the opening, it may be necessary to partially or even wholly close it, for it is difficult to devise a covering for the exterior which would baffle every wind that blows.

The market is so full of all kinds of appliances called "Ventilators", that **difficulty in selection** is experienced by those who have not a clear conception of what is required, or of what is possible of accomplishment thereby. The evil is, that so many people have a vague idea, fostered by the claims of rival manufacturers, that, when ventilators are once fixed, constant and comfortable ventilation can be secured without further personal attention. If it be clearly understood that such is impracticable, then their employment will not be discontinued, but they will be used with greater discrimination and regulated to suit the varying changes of weather.

Fig. 568. "Sheringham" Air-inlet.

The simplest form of air-inlet is the "Sheringham" (fig. 568), which necessitates a hole through an outer wall with a grating on the outside, and an inner frame provided with a hinged hopper-shaped flap, weighted so that when a weight attached to a cord is raised, the flap may be opened at will; this is an useful appliance when suitably placed and regulated, but, as ordinarily used, too small and consequently liable to cause draughts.

Fig. 569 is similar to a "Sheringham", but made larger and provided with a pierced inlet-plate, which has the effect of causing a better distribution of the incoming air.

Fig. 570 shows a **Sheringham air-inlet, with baffle-plates**, and fig. 571 a similar inlet, but with a screen in the middle of the wall for breaking strong currents of air on windy days. These inlets can be cased inside the room with woodwork in the form of a bracket, as shown in fig. 572, if desired.

An inlet with a **regulating valve** placed within the thickness of the wall, instead of a hinged front, is shown in fig. 573; while fig. 574 illustrates a somewhat similar inlet, which contains in addition to the baffle-plate b and valve c, a water-drawer e intended to catch and retain the dust in the passing air.

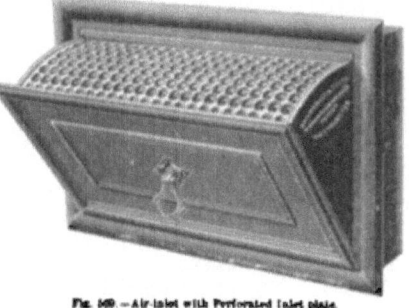

Fig. 569.—Air-inlet with Perforated Inlet plate.

A **drawer air-inlet** with baffle-plates is shown in fig. 575.

Fig. 576 is provided with **louvres behind an ornamental plate**, with means for

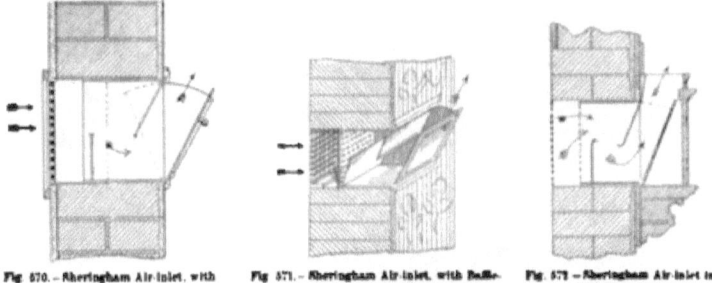

Fig. 570.—Sheringham Air-inlet, with baffle-plates.   Fig. 571.—Sheringham Air-inlet, with Baffle-plates and Inner Wind-guard.   Fig. 572.—Sheringham Air-inlet in Wood Bracket.

regulating the louvres. The same arrangement, but with the louvres sloping in the opposite direction, is made to act as an outlet.

Fig. 577 shows a small appliance for fixing over a slit in **the bottom rail of a window**, and capable of being closed; it may be employed when windows cannot be opened, but can only admit a limited amount of air as a supplement to other openings. Hit-and-miss gratings for fixing in the meeting-rails of a sash-window can also be obtained. A deep bottom rail and bead to an ordinary double-hung sash permit of the window being opened to allow air to enter at the meeting rails.

These illustrations have been selected to indicate the many types of air-inlets now in the market. It will be observed that the object generally aimed at is the admission of air, so as to cause diffusion throughout the apartment, and to avoid a direct current or draught; but, as previously stated, they

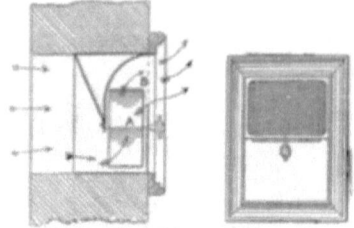

Fig. 572.—Air-inlet with Regulating Valve A, and brass gauze front B.

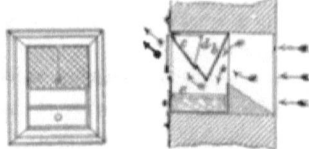

Fig. 574.—Air-inlet with Baffle-plate b, valve c and water-drawer a.

are frequently employed too small in size, and improperly located, so that, when the suctional force within is powerful, they either fail in those particulars for

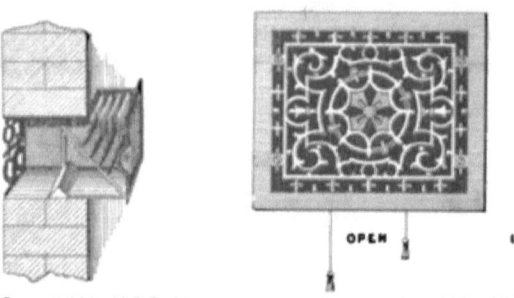

Fig. 573.—Drawer Air-inlet, with Baffle-plates.

Fig. 576.—Air-inlet with Movable Louvres.

which they have been designed, or else do not give an efficient supply of air, the result being draughts, or air being drawn from other sources.

Fig. 577.—Air-inlet for Fixing on Bottom Rail of Window.

Rotary air-propellers or fans are now acknowledged to be the most economical appliances for impelling air into buildings when the plenum system is adopted. Various manufacturers have introduced different forms, the differences being principally in the form and number of the blades. Fig. 578 shows one with two blades, and fig. 579 shows one with six blades of a different shape. As, however, the plenum system, for which they are principally required, will not be fully discussed for reasons previously given, it is unnecessary to consider closely the several points of difference they present, or to go further into particulars respecting their application.

The use of a spray of water for inducing a current of air in an extract-tube

Fig. 578.—Haworth's Two-bladed Air-propeller or Fan.

Fig. 579.—Baird, Thompson, & Co.'s Six-bladed Air-propeller or Fan.

will be considered and illustrated in the next chapter; exactly the same arrangements can be adopted for inlets.

Air-inlet cowls, such as those shown in figs. 580 and 581, are sometimes used; they are useful for spaces surrounded (or nearly so) by other rooms, such as halls

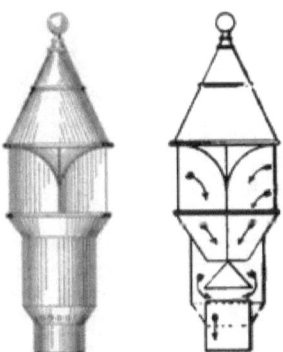
Fig. 580.—Elevation and Section of Donald & Sime's Air-Inlet Cowl.

Fig. 581.—View of Baird, Thompson, & Co.'s Air-Inlet Cowl.

and staircases, as well as in ordinary circumstances. They must be fixed on roofs where the wind will have free play upon them, as they will otherwise act as outlets. In still weather also they will cease to serve the purpose for which they are intended.

Warmed-air inlets will be considered in Chapter VIII., under the head of "Warming and Ventilation".

CHAPTER V.

AIR-OUTLETS.

Too much stress cannot be laid upon the fact that in most British homes, where open fires are provided in almost every room, **the smoke-flues from the fireplaces** must be regarded as the exits for air. Under ordinary circumstances they are sufficient for the purpose; when a fire is burning in the grate, the smoke-flue is practically the only way whereby the air of the apartment finds its way to the open. It is therefore useless to provide other openings as exits, without the employment of heat or mechanical power, because such will almost invariably be found to act as inlets and not as outlets, and will be liable to cause discomfort by admitting too directly large volumes of cold air in improper directions.

To place **a mica-flap ventilator** (figs. 582 and 583), as an outlet, in the external wall of a room with an open fire, is a very common mistake; when the windows and doors are closed, the suction up the flue will almost invariably close the flaps, so that air cannot possibly escape in that direction, and air will rarely force its way out past such flaps, even when doors and windows are freely opened. The only useful position in which a

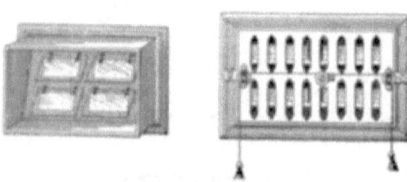

Fig. 582.—Mica-flap Air-outlet.

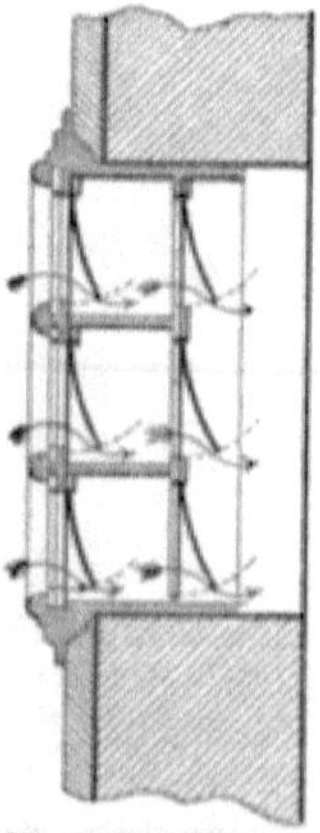

Fig. 583.—Section of Mica-flap Air-outlet with Double Flaps.

flapped ventilator can be fixed is in direct connection with the flue from the fireplace, towards the upper part of the room; then, provided the throat of the flue is throttled and there is a good draught upwards, the heated air from the upper portion of the room will be carried off. This, however, is only an advantage when a room becomes overheated, and an objection to ventilators so placed is the almost certain disfigurement of the walls and ceiling by smoke being at times drawn into the room. Better ventilation of the apartment could be secured by the employment of a suitable and well-placed air-inlet, with the fireplace flue acting as the sole outlet.

If it is really considered desirable to employ a flapped ventilator, one has been devised which works within a segment of a circle, so as to avoid noise from

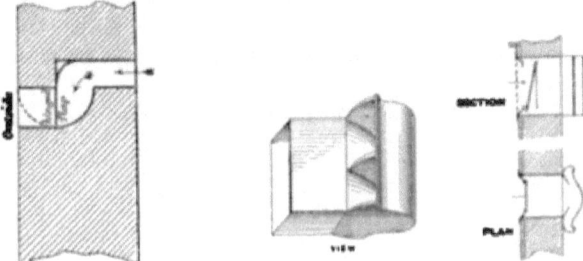

Fig. 584 — Improved Mica-flap Air-outlet.   Fig. 585 — Projecting Air-outlet for Fixing in Walls.

the movement of the flaps, whenever the forces within and without vary. This form (fig. 584) is, however, unsuited for connecting with smoke-flues, because the flaps can never be made to fit so tightly that smoke will not pass around the edges; consequently its use is very limited.

For rooms without fireplaces and ventilation-flues, **outlet ventilators in the external wall** may be of service. A ventilator of this kind is shown in fig. 585; it can be supplied with a valve, as

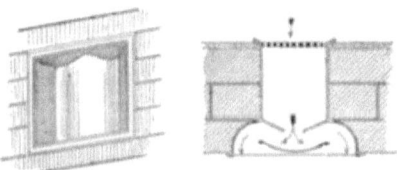

Fig. 586 — Flush Air-outlet for Fixing in Walls.

shown in the plan and section, to prevent any inrush of air. Another form is shown in fig. 586, which has not the disadvantage of projecting beyond the face of the external wall, but which may not be absolutely free from back-draught.

Extract-cowls on outlets, such as are shown in figs. 587 to 597, are designed (a) to act as wind-bafflers so as to obviate a direct blow down the outlet-shaft, (b) to prevent the inlet of wet, and (c) to accelerate the outflow of air by the suctional force of wind acting upon the cowl.

Rival claims by numerous manufacturers have caused experiments to be made with a view to demonstrate the relative value of the various forms employed, and the conclusion arrived at,—after many trials carefully made by Commissioners appointed both in England and in America,—is that **an open tube, with a protecting cap** to prevent the entrance of wet, gives as good a result as any of them, if varying states of the weather are taken into account. When a strong wind is blowing, change of air can, as a rule, be secured within a building without special means, and to add to the facility of obtaining change of air at such times may simply mean greater discomfort to the occupants. In a still atmosphere, when change of air would be of advantage, the cowl exercises no power whatever, and unless provided with a valve, as in fig. 597, there will certainly be a down-draught instead of an up-draught, if a suctional influence, such as an open fire, be employed within.

Nevertheless, **outlet-cowls may be usefully employed** in connection with

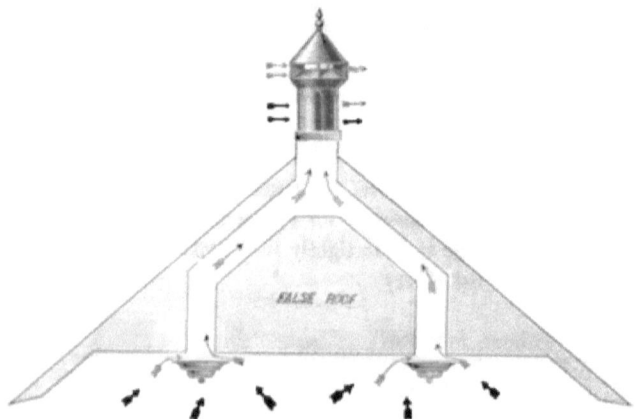

Fig. 597.—Walker's Extract-cowl, with Two Shafts.

buildings in which there are no open fires or other upcast flues, provided there are also suitable inlets. In many buildings—such as schools, churches, chapels, public halls, &c.—which are only occasionally occupied, changes of air can thereby be effected at times when no one is present. Such changes of air will greatly assist in keeping the buildings in a pure and healthy state for occupation when required. Means, however, should always be conveniently placed for regulation, or even closing both the inlets and outlets; otherwise, in cold weather, discomfort from draughts will be experienced. There should, as a rule, be only one outlet to an apartment, suited to the size thereof, for if the outlets be multiplied, particularly when fitted with cowls, one may draw upon the other at times when the action of the wind is irregularly exercised upon them. Several

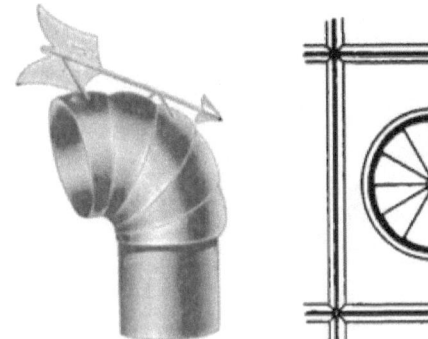 

Fig. 588.—View of Lobster-back Rotary Cowl.   Fig. 589.—Rotary Ventilator fixed in Window-pane.

openings from the apartment connected to one cowled outlet, as in fig. 587, may be advantageously used for withdrawing air from different parts of the same room.

The different kinds of extract-cowl are very numerous indeed, but they fall into two distinct classes:—(1) Cowls with movable parts; and (2) Cowls without movable parts.

(1) **Cowls with movable parts** are made both for inlets and outlets. The "lobster-back" rotary cowl, shown in fig. 588, is an example in point. If arranged so as to expose the open side to the wind, it will act as an inlet, if the reverse as an outlet; consequently its power in either direction is in proportion to the external force of the wind, provided no other power is exercised from within. Should the latter take place, one may either counterbalance the other, or, what is more probable, the current will vary in proportion as to whether that within or that without is the stronger.

Rotary ventilators, such as those shown in figs. 589 and 590, are often attractive because they give ocular demonstration that air is moving, but without the accompaniment of mechanical force they are most delusive. Fig. 589 is a form at one time employed, fixed in windows or doors, but now rarely seen. The revolving of the blades doubtless has the effect of diffusing the air, but such can be accomplished with better results, as previously indicated, without any movable parts.

Fig. 590.—Howorth's Archimedean Screw Revolving Ventilator, with part removed to show the Screw.

When rotary ventilators, such as fig. 590, are used as outlets, they impede the flow instead of assisting it, unless wind is blowing with sufficient force to revolve the outer louvred cap; consequently the discharge is in varying pro-

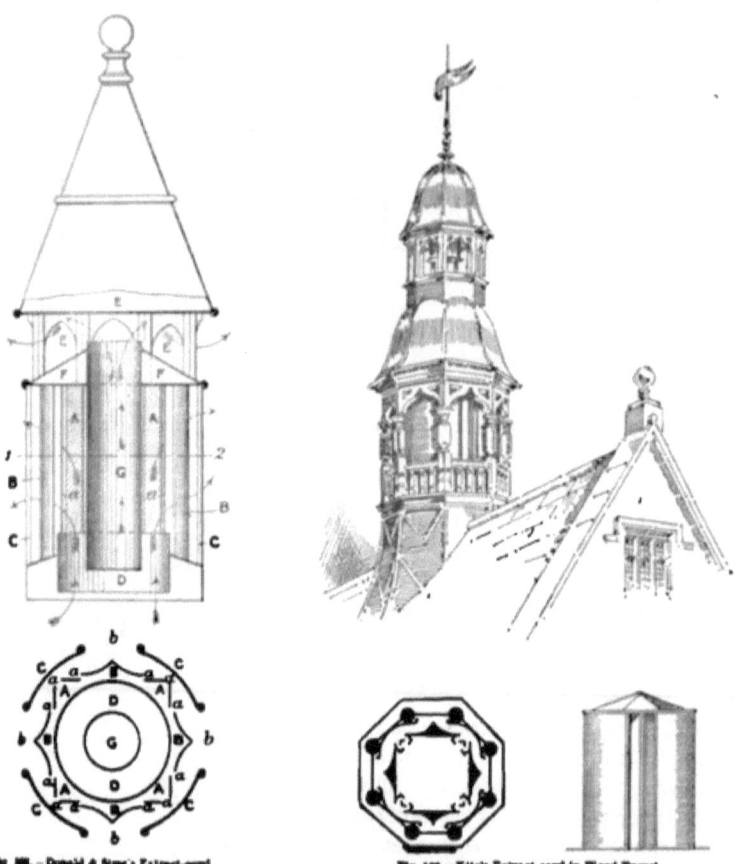

Fig. 591.—Donald & Son's Extract-cowl.

Fig. 592.—Kite's Extract-cowl in Wood Turret.

portion according to the power of the wind outside, and the cowl would probably be equally effective without the revolving parts. Some people suppose that down-draughts are prevented by this form of ventilator, but much depends upon the direction in which force is exercised; if from within, (say) resulting from an open fire in an apartment unprovided with other sufficient inlet, the tendency will be to suck air in past the screw, when the external air is comparatively calm; and, if wind be blowing, antagonistic forces may be brought into play, one being the suctional force exerted in consequence of the fire, which will endeavour to draw air in, the other being the force of wind acting on the louvred cap by which the screw will be revolved and made to draw the air out, the result of which might be to reverse the action in the flue, and cause smokiness within.

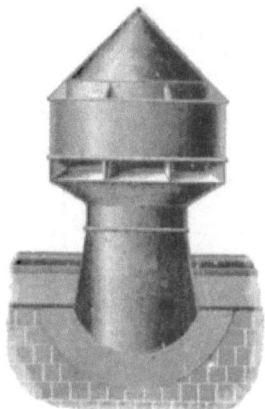

Fig. 593.—Bedford's Exhaust ventilator.

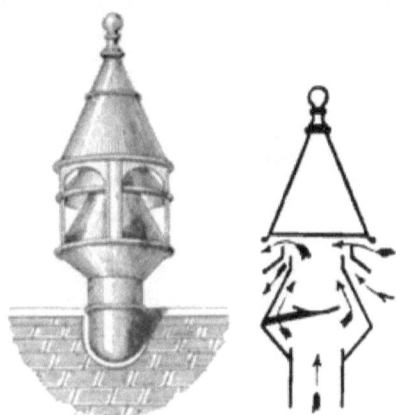

Fig. 594 — Donald & Sims's Extract cowl.

When turned by mechanical power, however, rotary ventilators may be usefully employed to extract air from apartments in which there are suitable inlets and no open fires.

(2) **Cowls without movable parts** are preferable, where mechanical power is not available. Speaking broadly, we may include in this category not only those circular, octagonal, or square cowls, upon which winds from every quarter take effect, but also the rectangular dove-cot ventilators with blank ends, and what are known as "concealed roof ventilators".

The most common form of extract-cowl is the *circular cowl, with vertical plates and openings* so arranged that the wind, passing over and between the plates, sucks air out through the openings, and so through the central shaft, with which these are connected. The arrangement of the several parts by different makers differs considerably, but the general principle remains the same. Fig. 591 illustrates a good cowl of this sort, and fig. 592 a somewhat different one, designed for inserting into a wood turret, as shown.

Cowls with the air-slits arranged horizontally instead of vertically, are also successful. Fig. 593 shows a good example, which is fitted with condensation-channels inside and outlet-holes, and with internal valves for opening and closing. A simpler form is shown in fig. 594, and another in fig. 595, the latter having a rather curious appearance.

Fig. 596 shows a cowl, in which there are *both horizontal and vertical openings*, but experiment alone can indicate whether this modification increases the extracting power. See also fig. 587.

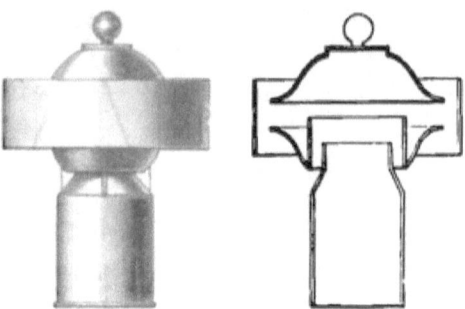

Fig. 596.—Benton Gibbs Extract-cowl.   Fig. 596.—Walker's Extract-cowl.

A *different arrangement* is shown in fig. 597; in this, the wind passing horizontally through the upper part draws the air up through the holes in the perforated plate, while back-draught is prevented by the light mica disc, which is hinged on one side. In another cowl, made by the same firm, the mica disc is arranged to move up and down a central rod, and this arrangement appears preferable. In either case, the weight of the disc must be overcome before extraction can take place.

The *dove-cot type of extract-ventilator* is shown in fig. 598, and a modified form in fig. 599. These are not as universally successful as the cowls previously described, but in some situations their appearance may be preferred.

"*The concealed roof ventilator*" was designed to meet the wishes of those to whom all projections of ventilators above the roof are eyesores. A good ventilator of this kind, having outlets for moisture, is shown in fig. 600. These ventilators are not as efficient as good cowls, but they have their uses.

Flue-outlets, as shown in fig. 601, may assist as outlets when there are no fires lighted, but when the fires are used they will almost invariably become inlets, and consequently at such times should be closed. For double chimney-stacks

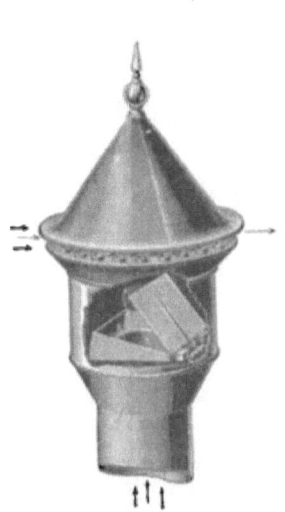

Fig. 597.—Sang's Extract-cowl.

Fig. 598.—Kite's Extract-ventilator of the Louvre-cut Type.

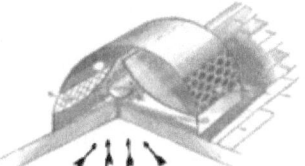

Fig. 599.—Baird, Thompson, & Co.'s Extract-ventilator.

they must be placed in pairs, the wind blowing right through both ventilators. "**Water-spray ventilators**" have been used with considerable success. In

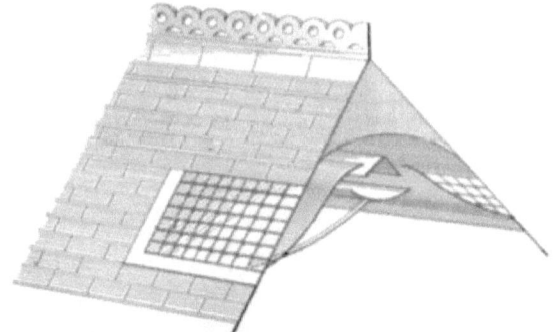

Fig. 600.—Donald and Sims's Concealed Roof Ventilator.

these ventilators the force exercised by a spray of water within a suitably-constructed tube is employed for inducing a current of air either into or from an apartment. Where water at sufficient pressure is available, these ventilators may be usefully employed. Their application, however, must necessarily be very limited, because of the cost involved in many localities, compared with the result effected, which in most instances can be secured by far simpler means. In cold weather there is also the liability to freezing of the water. Fig. 602 illustrates an extract-ventilator of this kind, in which the water-spray from the nozzle A draws the air vertically downwards. Other appliances are made in which the spray acts vertically upwards, and others again in which it acts horizontally.

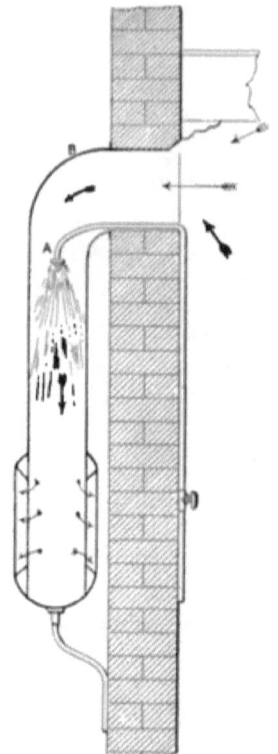

Fig. 601. Kite's Chimney stack Ventilator.

Fig. 602.—Section of Kite's Water-spray Vertical Exhaust, driving downwards.

CHAPTER VI.

"NATURAL" VENTILATION.

Much has been said and written as to the respective advantages of **upward and downward ventilation**, and I am inclined to think that the more general condemnation of downward ventilation has been arrived at without full consideration of the facts. A statement previously made—that the flue from a fireplace is almost the only (if not the only) air-outlet from most rooms—can easily be verified by any one, particularly when a fire is burning in the grate. Open a window, or other "ventilator", wherever it may be about the room, and it will be found that *air will as a rule enter*. In fact, only when the suctional power of the wind outside can overcome the draught up the flue, will air leave the apartment in any quantity except up the flue, provided of course that the door is shut, so that other influences within the house may be excluded. Inasmuch as fireplaces are invariably placed near the floor, and in all modern fireplaces the lowest point of the flue is generally less than three feet above the floor, extraction of air in such cases must be in a downward direction from the ceiling towards the fireplace, unless inlets occur in improper positions,—*e.g.* about the lower portion of the room,—when quick horizontal movements of cool air direct from the inlet to the fire (in other words, draughts) may be set up, to the discomfort of the occupants.

There is nothing more important in the attainment of comfortable ventilation than to estimate rightly **the functions of the fireplace flue**. Most smoke-flues are built about 14 inches by 9 inches, and, when pargetted, will measure 120 square inches; but often at the throat, and more frequently by the chimney-pot, this area is reduced to between 60 and 80 square inches, or (say) half a square foot. As the lengths of flues vary considerably, and the question of no fire or more or less fire in the grate exerts a varying influence upon the amount of air which in a given time will pass up the flues, it will be seen how greatly must differ, even under similar conditions of the external atmosphere, the rate at which the air of various apartments will change, and, as the conditions vary outside, greater differences still will be found in the amount of change of air that will take place within the various rooms. Nevertheless, in any apartment provided with an open fireplace, the smoke-flue is the dominant means for securing an outflow of air.

It is possible to vary the size of the flue opening by means of **a register**, an appliance which fortunately is now less employed than formerly, because it is, as previously mentioned, too often wrongly used to entirely close the flue, and it is only in the smallest rooms and in cold weather that the ordinary smoke-flue can be considered too large for the purpose of ventilation.

Regulation of the amount of the change of air in an apartment is better effected at the inlet than the outlet; consequently all inlets should be capable of easy regulation. They should also be of larger area than the outlet, viz. the fireplace flue.

Windows and doors should in every dwelling be regarded, in addition to their other functions, as definite appliances for ventilation. Generally they serve as inlets for air, but at times as outlets. When an open fire is burning in a room, they are almost invariably inlets only, and in the majority of rooms in which there is a fireplace, windows and doors are by far the most effective appliances which can be employed for ventilation, provided they are suitably placed, and can be easily opened, more or less, at the discretion of the occupants. Although it may not be comfortable to have windows and doors constantly open when the rooms are occupied, there are many times when they may be partly opened, and on other occasions, when the rooms are unoccupied, they may all be thrown open, so that a thorough change of air may take place, regardless of draughts.

In stating this, there is no intention of condemning as useless **other inlet-appliances** when intelligently fixed and regulated, but there is a danger, which many people fall into when such appliances are provided, of trusting to these alone, and making them the excuse for not freely opening windows and doors; whereas all such appliances, which necessarily cannot be of very large dimensions, ought only to be regarded as supplementary to the open windows.

Doors have been mentioned above as appliances for securing change of air to an apartment; as such they are frequently and usefully employed, but a distinction must always be made between outer doors and those to the separate rooms, and it must be ascertained whether at the time they serve as outlets or inlets. If change of air takes place by way of an inner door, it must be noted that, should air pass out of the room thereby, it will probably proceed thence through the hall and by the stairway into some other rooms, the doors of which may be open, so that, if the air has become fouled in the first room, its foulness will become in part communicated to the hall, staircase, and other rooms; and when air enters a room by the door, it comes from other parts of the house, and the question will then arise whether it has in any way become contaminated in its passage.

The position of the door of a room, and the side upon which it is hung, will have a considerable effect upon the comfort of the occupants of the room, because, even if each compartment of a house be separately supplied with a suitable air inlet and outlet, the balance of power is so sensitive that, as the door is one of the largest openings round a room and the one most frequently opened and closed, it must exert considerable influence upon the air movements within the room; it must therefore be regarded as a ventilator. Fig. 603 indicates various positions and directions in which doors may be hung, the best being those in which the incoming air has the greatest distance to travel from door to fireplace.

The advantage of securing change of air through the doorways is that, in cold weather, the air is more likely to become tempered by passing through portions

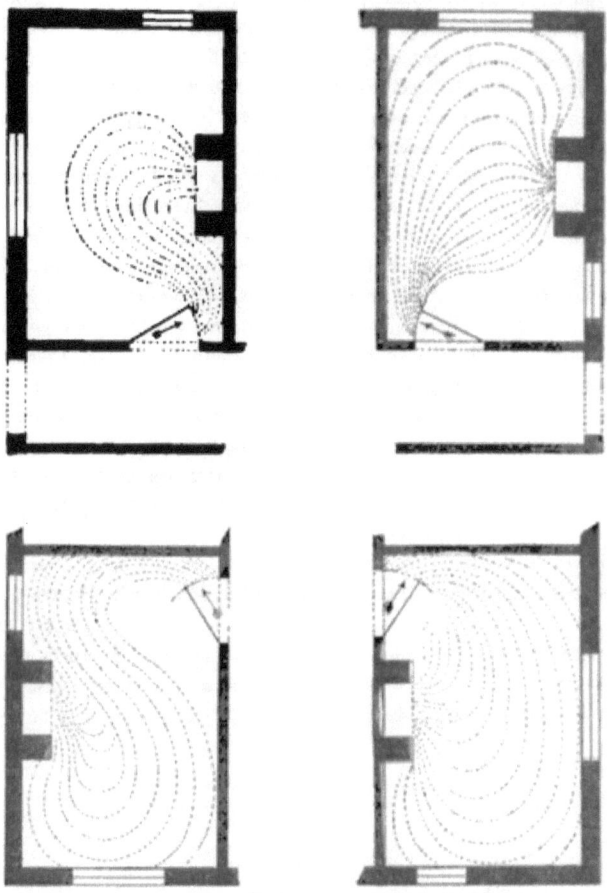

Fig. 613. Four Arrangements of Doors and Fireplaces, showing Currents of Air when the Doors are open

of the building, and the change is consequently less unpleasant to the occupants than when air is allowed to enter direct from the exterior. The air can also be easily regulated in quantity by opening the door more or less, but as the privacy of the room is thereby lessened, such an expedient is not at all times convenient;

the difficulty can, however, be overcome by the use of fan-lights over the doors, made to open, as shown in Plate II.

The ventilation of all rooms in a house through the hall and stairway has been advocated, and openings have in some instances been provided above the doorways, fitted with hinged casements or with flaps or louvres, and if a good supply of fresh air be admitted to the hall and stairway, warmed in cold weather, and all chance of contamination within the house be avoided, such a system may be carried out with good effect. But when open fires are employed in some rooms only of the house, and the air-supply to the hall and stairway is inadequate, the stronger suctional power induced up the flues in such rooms will have a tendency to draw air from other rooms where there are no fires, perhaps down the smoke-flues, and may thus cause a sooty smell throughout the house; or where the staircase is of considerable height, the tendency is for the warmed air to rise to the top, and, if it can find exit there, it will cause a pull upon the air of the apartments instead of supplying air thereto. In fact, there are many houses not specially arranged to be ventilated as above described, in which this tendency of a lofty stairway to act as an upcast shaft and draw air from the rooms, becomes most troublesome, and is a frequent cause of smoky chimneys, or at least of down-draughts in flues where there are no fires alight. Consequently it is, as a rule, better to provide separately for the ventilation of each room, as well as of the halls and stairways.

Suggestions have been given how this ventilation may best be accomplished, but **it is impossible to lay down absolute rules** which will in every case be effective. Conditions are so widely different that, to obtain success, each case must be treated on its merits, and little more can be done than explain the salient points which should demand attention, and ever to insist upon the necessity for intelligent regulation if the comfort of occupants is to be secured.

CHAPTER VII.

CONTAMINATION OF AIR, &c.

The principal causes of the contamination of air in rooms, other than want of adequate change of air, are: (1) *Deposits of animal and organic matter, including exhalations from human beings and animals, the products of respiration, &c.*; (2) *Condensation*; (3) *Evaporation*; and (4) *Absorption*.

In every inhabited room, particularly those in which meals are partaken of, there must necessarily be deposited a certain amount of **organic matter**, such as crumbs, particles of skin, respiratory matter from human beings and domestic animals. This matter, if permitted to accumulate, will putrefy, and give off gases, which will pollute the air and make it unwholesome to breathe.

On all surfaces, which by not absorbing and retaining heat are below the normal temperature of the room, **condensation** will take place, particularly when change of air is not frequent. In addition to the moisture or vapour of water, which is ever present in the atmosphere, moisture is caused by respiration and exhalations from human beings and animals, steam from cooked meats, the products of combustion (whether for heat or light), tobacco-smoke, and the volatilization of various organic substances; every one of these forms of moisture, with their several organic or other impurities, may be condensed upon the surfaces within the room, when these are below the general temperature of the atmosphere it contains and the change of air is not sufficiently rapid to at once carry them away.

This condensation can always be noticed upon large surfaces of glass, when the temperature outside is below that within, but although it is more noticeable upon the glass, it is, in fact, being deposited upon all the surfaces in the room in proportion to their temperature; hence the importance of constructing and furnishing dwellings with materials which absorb and retain heat, and not with normally cold and impervious substances.

Much of the moisture, and of the more volatile portion of the substances condensed, will again be carried off by **evaporation** when the temperature of the surfaces is raised, or a freer circulation of air is permitted, but the less volatile organic matter remains and forms a film over the surfaces, which will either putrefy and so contaminate the air, or will dry and fall off as dust, probably so fine that it will be carried about by movements of the air. In this way, the germs of disease may contaminate the air, and enter the lungs of persons occupying the room.

Condensation and evaporation, therefore, play an important part in connection with ventilation, and I am inclined to believe that the popular fancy for hard, impervious substances, which do not readily retain heat, for the construction and lining of walls, is more than questionable, if comfort and good ventilation are to be secured in our homes.[1]

[1] It must not be forgotten that those materials which "absorb and retain heat",—such as wool, ordinary plaster, unglazed wall-papers,—are the very ones to absorb organic matter. Sir Douglas Galton, in his book on *Healthy Dwellings*, says that "in 1862, in the French Academy of Medicine, a case was mentioned in which an analysis had been made of the plaster of a hospital wall, and 46 per cent of organic matter was found in the plaster". This was

Many substances and materials possess, in varying degree, **the property of absorption**, not alone of moisture but of volatilized substances and even of gases; for example, clothes and draperies retain for a considerable time the odour of tobacco-smoke. Even to plain surfaces, air itself somewhat clings, for it is a well-known fact that a stream of air, propelled at an angle against a flat surface, does not rebound at the same angle as would a solid, but goes off at a more obtuse angle, indicating that an attractive or retarding influence has been exerted.

Some, who have observed the stuffy effect produced by **a superabundance of draperies**, rush to the opposite extreme, denude their rooms of all such, strip the papers from the walls and paint them instead, or cover them with tiles and glazed ware, discard carpets, and varnish the floors. Yet, in doing so, they probably fall into greater error, for to secure change of air in such rooms, without causing draughts, becomes more difficult; the rooms have a chilling effect in cold weather, and unless the hard surfaces are constantly cleaned, they become incrusted with dirt resulting from condensation thereon.[1]

Observation has demonstrated that **the dust ever present in the air** influences the deposition of moisture, every particle of which carries with it an atom of dust; in addition to which, the moistened surfaces will cause other particles to become attached. Doubtless most dust-particles are inert; some, however, are organic, and of these some are living microbes. Hence it must be a questionable proceeding to give resting-place for such, particularly in conjunction with moisture, which is supposed to retain their vitality, and under favourable circumstances to assist in their multiplication.

Frequent change of air too often brings with it a large amount of dust and soot, &c., and the careful housewife will often close windows and doors and ventilators to keep it out. In doing so, ventilation may be retarded. Except in very windy weather, when heavier particles are raised and blown about, it is a question whether, in badly-ventilated houses, dust is not deposited upon

---

before the days of bacteriological examination, but undoubtedly this organic matter contained many micro-organisms, possibly pathogenic; even ordinary bricks have proved capable of harbouring and multiplying this minute life. Besides, absorbent materials do not prevent condensation; they merely hide it. The true remedy would appear to lie in the direction of warming the surfaces in a room, especially the walls, and of providing a freer circulation of air. An interesting example of a house constructed on this principle is given by Dr. Billings in his book on "Ventilation and Heating". The house is at Creil in France, and is of two stories, the external walls having an outer skin about 9 inches thick, and an inner skin 4 inches thick, with a cavity about 9 inches wide between. The air circulating in the cavity is warmed by means of hot-water pipes to a temperature of about 120° Fahr., and gives to the inner skin of the wall a temperature of about 90° Fahr. The air to the rooms is admitted cold, and is extracted by means of "chimneys". "The location of the house is a damp one, but the interior is dry, and the house is said to be very comfortable."—ED.

[1] On the other hand, so great is the absorption of organic and other impurities by ordinary wall-papers and plaster, that every sanitarian declares that old wall-papers must be stripped off before new ones are laid, and some go so far as to advocate the removal of the outer skin of plaster every few years.—ED.

articles of furniture, walls, floors, &c., in greater quantities than where an adequate movement of air is freely permitted. Naturally there may be various reasons why at different times more or less dust will be deposited in rooms, but, in a properly-ventilated room, the finer particles of dust will not readily accumulate.

The ground on which the house is built is often a fruitful source of contamination to the air within. Not only excessive moisture or damp, but also noxious gases, may be drawn up from the ground by the warmth of the house. Hence the necessity[1] of draining the subsoil, of rendering the basement-walls as impervious as possible, of covering the site with an impervious ground-layer, and of laying on all walls, immediately above the ground, a thoroughly damp-proof course.

A basement story under living-rooms is generally an advantage, provided it is kept dry and clean, and is well supplied with fresh-air inlets and up-cast flues. In winter it often happens that, when the house is closed, an adequate amount of air is not supplied to the several apartments, and when fires are lighted, the basement-air is, to a large extent, drawn upon to take the place of that which ascends the flues. A fusty smell within a house is often traceable to dampness, or to an accumulation of rubbish, in the basement. To remedy such a state of things, the basement should be cleaned, dried, and provided with suitable air inlets and outlets, and at the same time separate air-inlets should be constructed to each room in the house, with the fireplace flues as exits.

Where there is no basement, and a space is left between the ground and the floor immediately above, equal care should be taken to keep the space dry and clear of all contaminating influences, because when windows, doors, and ventilators are closed, such spaces will certainly be drawn upon to supply air to the rooms above. Many people would be surprised to find, if the floor were removed, what an amount of filth there often is in such places, and they might then realize why the rooms are close and stuffy, and why unhealthiness has frequently occurred among the inmates. When such air-spaces are properly constructed and cleared of all rubbish, it is still necessary (in order to prevent the decay of wood and the stuffiness alluded to) to provide for change of air therein, and the height of perfection would be to do so quite independently of the rooms above, but such is rarely practical. A separate flue or flues carried up with the smoke-flues, and air-inlet gratings suitably placed around the walls, will generally suffice.

Foul gases from drains and waste-pipes are a dangerous kind of air-contamination, and particular attention should be given to the principles laid down in

---

[1] As explained in Section II., vol. i. pp. 54, 69-71, and 75-85.—ED.

the sections on "Sanitary Plumbing", "Sanitary Fittings", and "Drainage", in order that the house may be free from all these deadly emanations. Even when all waste-pipes discharge over trapped gullies, particular care should be taken that the air-inlets to the rooms are as far from them as possible, as occasionally the gullies may choke and overflow, or the standing water in the traps may evaporate and allow the free escape of air from the drains.

Faulty construction of the house, as explained in Chapter III., is another source of air-contamination, and in this connection the following paragraph from Section II. (vol. i. page 61) may be quoted:—

"Since dirt is so prevalent, it behoves the architect to avoid as far as possible all ledges, nooks, and crevices, and all unseen spaces which could give it lodgment. Considered in the light of cleanliness, the ordinary floor, with its plastered ceiling below and gaping boards above, is radically wrong; so also is the confined space so often provided between the ground-layer and ground-floor; so also are lath-and-plaster partitions, hollow walls, and indeed all details of building-construction which provide space invisible and inaccessible to the householder. Sooner or later dirt finds its way to these dark places, and vermin breed and wander there, safe from the housemaid's broom and the cat's eager paw."

Free admission of daylight to enclosed spaces intended for occupation appears to be necessary for maintaining them in a condition suitable for healthy residence, and therefore light is an important factor in connection with effective ventilation. Its purifying action upon the atmosphere is most marked, and consequently should be encouraged.

Until the introduction of the incandescent electric light, air-contamination resulted from all methods of **artificial illumination,** these artificial lights being the result of combustion, by which complex impurities are thrown off into the atmosphere, carbonic acid predominating.

Candles, particularly those of common make, will, light for light, contaminate the atmosphere more than most illuminants.

Animal and vegetable oils come next, and then coal-gas. In oil-lamps, very much depends upon the form of lamp and burner, and the currents in the air of the apartment. Irregular and rapid movements in the air cause smokiness and imperfect combustion, which result in contamination of the atmosphere by products which are injurious to health when breathed.

The introduction of better illuminants has tended to a demand for more and more light, and the greater the ease and economy with which it can be used, the larger the quantity consumed. At the beginning of this century, a single candle would be considered sufficient for a room of moderate dimensions—later, an oil-lamp. Then probably three gas-burners would not be thought excessive, and now that people are generally becoming accustomed to the electric light, even more gas-jets may be demanded where the electric light is not available. All this continually-increasing demand for more powerful illumination has had a marked effect upon the ventilation of dwelling-rooms.

The introduction of coal-gas resulted in the demand for lofty rooms, which in turn require the use of more gas for lighting them, and yet in but few cases have additional inlets and outlets for causing increased change of air been provided; consequently, with a larger capacity and no better means of causing change of air, ventilation cannot be so good as in rooms less lofty,—*i.e.* of less cubic capacity,—because change of air in the latter can, under similar conditions, be more quickly brought about.

Various methods for conveying away the products of combustion, when gas is employed for illuminating purposes, have been adopted, but they cannot as a rule be relied on to act in a satisfactory manner under varying conditions of the outer atmosphere, and are frequently affected by the lighting of a fire, and the opening or closing of a door or window. Sunlight-burners have been extensively used, but the great heat emitted therefrom precludes their use in any but lofty apartments, and in them an excessive consumption of gas is required to secure the requisite illumination. Moreover, they act as a powerful means by which air is extracted from the apartment, and unless the greatest care be exercised in providing suitable inlets, they will cause down-draught in fireplace flues; and in cold weather it is difficult to regulate the inflow of air, required to replace that which they are the means of removing, in such a manner that discomfort may not be caused to the occupants of the room. The incandescent gas-burners now largely employed give an excellent light with less contamination of the air than in the case of ordinary burners..

The incandescent electric light has fortunately been so far perfected and cheapened, that it is placed within the reach of many who desire a strong illuminant without contamination of the atmosphere. An additional advantage which it possesses is the comparatively small rise in temperature involved, compared with the illuminating power, so that the ordinary means provided for securing ventilation are not materially affected when the electric light is turned on. It is in fact an ideal illuminant in connection with means for securing efficient ventilation, because it neither contaminates the air nor materially influences its movements. The methods of generating and applying it will be described in a subsequent section.

Wherever change of air in an inclosed space, such as a room, is retarded, and there are contaminating influences, **it is not alone the air which is vitiated,** but every surface therein becomes more or less tainted, and the longer such contamination takes place, the longer will be the time required (during which more frequent change of air is necessary) before such space and the articles within it will again become pure and the room rendered suitable for healthy occupation. Hence it is that every careful housewife realizes the necessity for the frequent removal of everything which is likely to cause air-contamination in and about her dwelling, as well as for the frequent airing of unoccupied rooms by throwing wide open the windows and doors.

Where there has been infection or continued contamination, it may be advisable to employ **powerful disinfectants,** not, as a rule, for making the atmosphere itself wholesome, but more generally for purifying the surfaces, objects, and parts of a building inaccessible to any other cleansing influence. For this purpose volatile substances are employed, which should be made to penetrate all cracks and crevices, ample time being allowed, so that the thorough destruction of what is injurious may take place. Fluid disinfectants may be employed for all accessible surfaces which would not be injured thereby. Many so-called disinfectants possess a strong odour but little antiseptic power; they are worse than useless in connection with ventilation, because they mask bad odours, which otherwise might indicate the sources of contamination and lead to their removal.[1]

CHAPTER VIII.

WARMING AND VENTILATION.

For some six months of the year in this country **artificial heat** is a necessity in our homes. At such times the outer atmosphere is too cold for the comfort of occupants when little exertion is being made. Various methods are consequently employed for raising the temperature of dwellings, such as open fires, closed stoves, gas-fires, gas-stoves, hot water, steam, and hot air. These systems of warming houses are dealt with in the Section on "Warming", and therefore they will only here be referred to as regards the effect they may respectively have on the ventilation of houses.

1. **Ventilating Fire-grates.**—The open fireplace has previously been referred

---

[1] For further information on this important subject see the "General Introduction", by Dr. Andrews, pp. 20-23, Vol. I.—ED.

to and the value of smoke-flues discussed, but in addition to the simple open fire, which undoubtedly is cheerful and pleasant under ordinary circumstances, although wasteful and frequently the cause of cutting draughts, there are more complicated forms specially designed to secure comfortable ventilation, the idea being to impart to the necessary incoming air, which must replace that ascending the flue, a portion of the heat given out by the fire.

Theoretically the idea is excellent, but unfortunately the difficulties which have to be overcome in the construction of the grate are considerable, and, although I have tried many forms, I **cannot say that any are permanently satisfactory**, the principal reason being that metal (generally iron) is employed, which must be jointed, and the constant variation in temperature alternately expands and contracts the parts, so that in time the joints open, and smoke is drawn into the fresh-air channels, vitiating the atmosphere of the apartments. One common fault they nearly all have, viz. the passage of cold air abstracts heat from the hot gases, given off at the top of the fire, to such an extent that soot is quickly deposited, and the temperature of the flue is proportionately reduced, frequently resulting in a clogging of the mouth of the flue and a lifeless fire.

I have obtained better results by utilizing the heat of the body of the fire, instead of abstracting it from the gases given off. Nevertheless I am inclined to think that in rooms of moderate dimensions it is better to be content with the radiant heat of the fire, in a well-constructed grate in which the fire is well surrounded with firebrick, and to admit a reasonable amount of fresh air by suitably-placed openings in the upper portion of the room as previously described; a further reason in favour of this opinion is that, as the air-inlet channels around the fire-grate cannot be easily got at and periodically cleaned, in time they must become fouled by the large quantity of air, which passes through and necessarily brings with it a considerable proportion of dust and dirt, and, to ground-floor rooms, the air-inlets are frequently too near the surface of the outer ground.

2. *Closed Stoves.*—Better heat-value in proportion to the amount of fuel used can be obtained from these than from an open fireplace, but generally at the expense of change of air, because the sectional area of the flue is less, and consequently the amount of air extracted is reduced.

Another result arising from the use of many closed stoves, particularly those principally constructed of iron, is that the metal becomes overheated; this causes a rapid circulation of air in the room, and large volumes of air are brought into contact with the overheated surfaces. Much of the dust floating about in the atmosphere consists of particles of organic substances; these either burn and give off carbonic acid, or are charred and cause an offensive odour. These heated surfaces also dry the air and make it unpleasant to breathe; to mitigate this evil, vessels of water should be placed upon or near closed stoves, and care should be taken to have the vessels and water clean.

3. **Gas-fires.**—If properly constructed, gas-fires are convenient and cleanly, and, as regards ventilation, might be expected to act in a similar manner to an open coal-fire, but in two important particulars they are less satisfactory:—

1. The heat from a gas-fire has a decided tendency to over-dry the air of an apartment, so that, unless a pan of water or other suitable appliance be

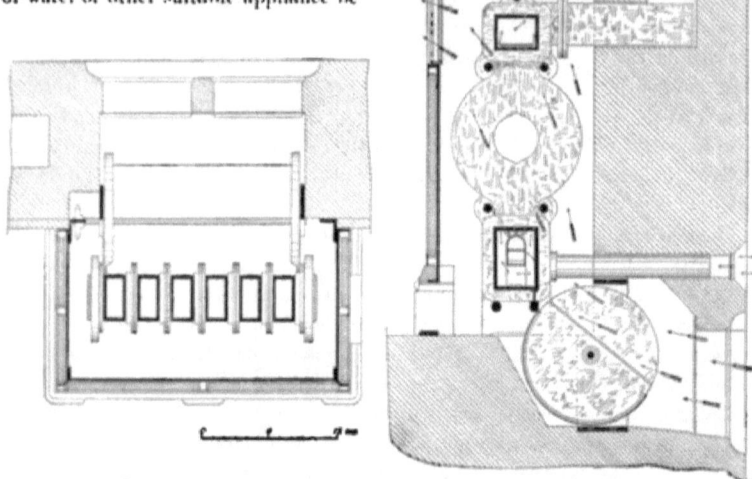

Fig. 604.—Plan and Section of Bruce's Single Apparatus for Warming Incoming Air by Gas.

used to supply sufficient moisture, the air becomes unpleasant if not actually unhealthy to breathe. 2. In setting grates for gas-fires it has become a custom to very considerably reduce the area of the flue, and in consequence change of air cannot take place so frequently within the room.

4. **Gas Stoves,** as far as our subject is concerned, may be divided into two classes, viz. those that are intended to assist ventilation, and those which make no attempt in that direction. As regards the latter, they should only be used when efficient means of ventilation can otherwise be secured.

Of those designed to assist ventilation, there are several varieties and forms, the object principally aimed at being to impart a portion of the heat, evolved from the burning of gas, to the incoming air of the apartment, and to convey the products of combustion into a flue or more directly to the open. A difficulty which the writer has experienced is to prevent back-draughts extinguishing the

Fig. 606.—Plan, Section, and End Elevation of Bruce's Double Apparatus for Warming Incoming Air by Gas.

flame, and thereby causing an escape of gas. Moreover, the working parts of such gas-stoves are principally of metal and have to be jointed; variation in temperature, arising from being at times highly heated and then allowed to cool down, causes varying expansion and contraction of parts, and ultimate breaking

218 VENTILATION.

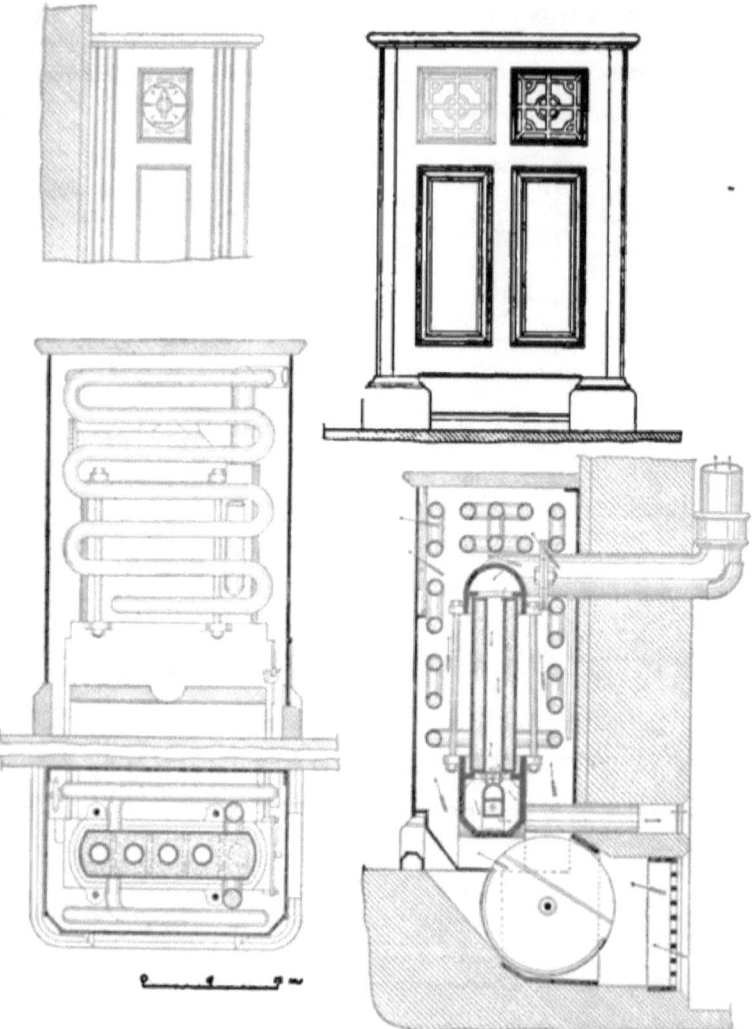

Fig. 606.—Plan, Sections, and Elevations of Bruce's Hot-water Apparatus or Radiator, heated by Gas.

of the joints, and the products of combustion also cause corrosion, so that any degree of permanence can scarcely be expected.

Improvements in arrangement and construction may be expected, and even now, in careful hands, there are several appliances which may be usefully employed. Of these, figs. 604, 605, and 606 are examples; they are well-designed and carefully-constructed ventilating gas-stoves, arranged so that the flame is fed from, and the products discharged into the open air, while fresh air is separately supplied to the room through the apparatus, and warmed thereby

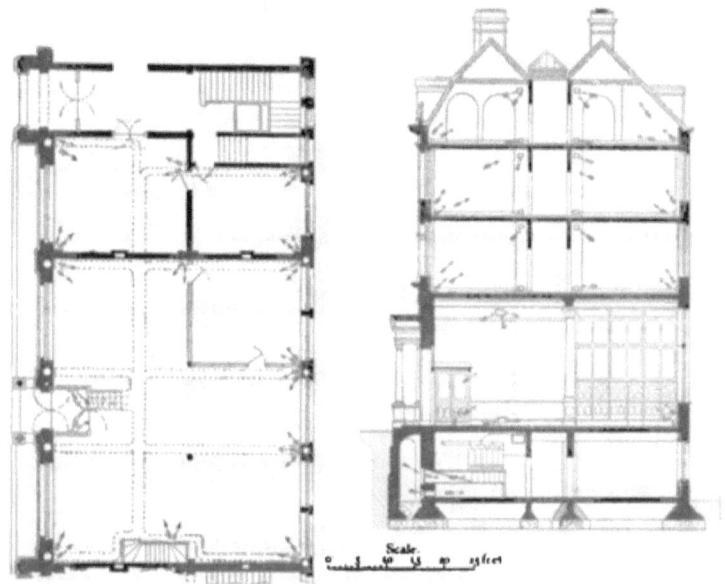

Fig. 607.—Plan and Section of Building showing Bruce's Radiators Warming the Incoming Air, and Upcast Shafts carrying off the Vitiated Air

when the gas is lighted. Fig. 604 gives the plan and section of a single apparatus: the outer case can be made of marble, faience, tiles, or mosaic, of any design or colour, in a metal framework. A double apparatus is shown in fig. 605; in this the front has a "projecting illuminating centre", in which three gas-jets are arranged to throw light through the stained-glass panels in the projecting portion; these gas-jets are perhaps more for ornament than use, although of course they will possess a certain amount of heating power.

In fig. 606 the plan, sections, and elevations of **Bruce's hot-water apparatus** or radiator with copper boiler are given, the water being heated by gas as in the preceding examples. The advantage of the water is that it will retain its heat for some time after the gas is turned off. Fresh air is admitted and warmed by passing over the coiled pipes.

Fig. 607 shows **Bruce's gas-stoves or radiators applied to the warming and ventilation of a building,** upcast shafts being provided for the vitiated air, and carried above the roof. This building, which, I understand, has been erected in Bradford, is an interesting experiment in the way of procuring warmth and ventilation by means of gas. In fig. 608 a scheme of warmed-air ventilation is shown, having a central air-warmer in the basement, and shafts therefrom conveying the

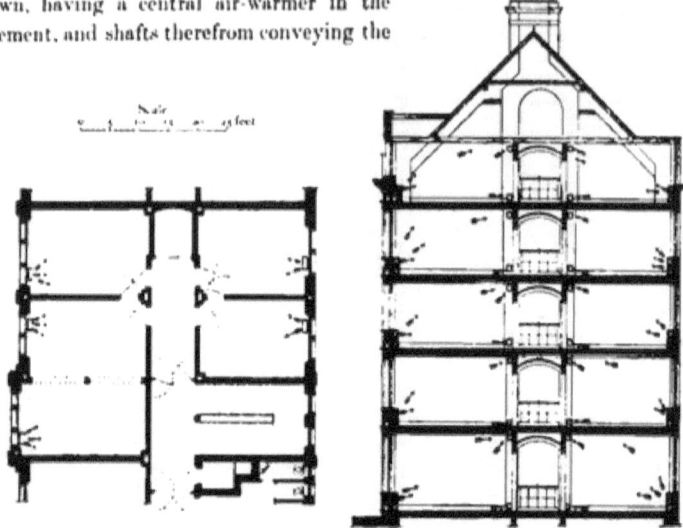

Fig. 608. Plan and Section of Building showing Bruce's Central Air-warmer and Shafts therefrom to the several Rooms, and Extract Shafts for the Vitiated Air.

warmed air to the several rooms. The vitiated air is carried away by upcast shafts as in the previous example.

With respect to these radiators, my fear is that, at times, the pressure of air on the outlet for vitiated air will cause a back-draught, which may extinguish the gas-flame, and that, with the outlets from the apartments arranged as shown in figs. 607 and 608, there will not be so thorough a change of air through them as there would be with the outlets on the same side as the inlets, and with the inlets and outlets all of larger size and with the latter near the floor.

5. Hot-water Heating.—In large rooms this is a convenient method of raising the temperature, but, unless supplemented by an open fire, has many of the disadvantages of closed stoves, and, unless special means for the ingress and egress of air are provided, good ventilation is impossible. Ventilating radiators have consequently been introduced,[1] by which air from the exterior is permitted to enter and pass through or around the tubes containing hot water; but even with that provision for admitting fresh warmed air, far too little attention is generally paid to the provision of exits, with the result that change of air does not freely take place.

Low-pressure hot-water heating is the simpler method, and when properly applied secures the greater comfort in rooms. High-pressure hot-water heating has found favour because the smaller pipes are less unsightly, but they have counterbalancing disadvantages. The small volume of water they contain, and the more rapid circulation set up in the pipes, results in greater variation in temperature, and when highly heated the air is made dry.

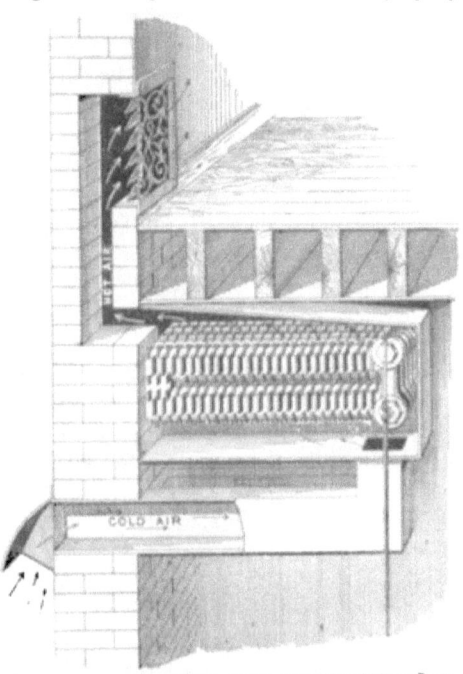

Fig. 608.—Steam Battery or Radiator admitting Fresh Warmed Air to a Room.

A **pernicious method of laying hot-water pipes** is to place them in trenches below the floor-level, covered with gratings, through which the sweepings of the floors find their way; the pipes thus become surrounded with filth, and when floors are washed moisture is added; the heat given off from the pipes facilitates decomposition, resulting in the contamination of the air of the room.

6. **Steam**, when directly used in place of water, adds to all these disadvantages, and should therefore be used with the greatest caution, if good ventilation is to be secured. Steam can, however, be more readily conveyed to greater distances than hot-water, and is therefore well adapted for large and extended buildings;

[1] For illustrations see Section on "Warming", pp. 126-7, Vol. II.—ED.

and when used in batteries to which fresh air can be admitted, such as that shown in fig. 609, ventilation may be assisted in cold weather, provided that suitable outlet-flues are properly arranged.

7. **Hot Air.**—Hot air is still more difficult to deal with effectively, unless mechanical means are employed for causing it to flow where required, and when motive power is made use of, it is far more effective and reliable when used for propulsion than for extraction, because, when a propelling force is employed, the source of air-supply can be ascertained and regulated; the air can not only be tempered, but may be cleansed and brought to a suitable hygrostatic condition, its volume can be determined, and its direction can be regulated at will. The means by which air is heated may considerably affect its condition. Low-pressure steam and hot water are far more reliable than stoves or furnaces, because the latter are liable to contaminate the air, particularly in the course of time, when joints open by frequent expansion and contraction, or by the wearing away of parts, and there is a further danger from fire should the air become overheated.

Whatever methods are employed when hot air is used, all **air-ducts and flues** must be readily accessible, and care must be exercised to keep them clean, otherwise in its passage through them the air will be contaminated and become unpleasant, if not actually unhealthy, to breathe.

Moreover, it is as necessary to have **suitably-arranged exits** as it is to provide inlets for the warmed air. They also should be kept clean, and, in order to avoid irregular currents and draughts, the exits into the open must be constructed so that movement of the outer atmosphere may exercise neither a suctional nor retarding influence upon the outflow.[1]

CHAPTER IX.

SUMMARY.

It has been my endeavour, in the foregoing remarks upon ventilation, to lead those who have to build and those who occupy houses, to realize the intricacies of the subject when regarded as affecting communities as well as individuals; that it is, in fact, the maintenance of the outer atmosphere in a state of purity,

[1] For further information on Combined Warming and Ventilating Systems see pp. 93–106, and 225–228, Vol. II., and Plates XX., XXI., XXII., and XXIII.—En.

without which pure air cannot be expected within houses; that contamination of otherwise pure air will take place in and about dwellings, unless constant cleanliness throughout is exercised; and that the situation and arrangement of houses, the materials of which they are built, the manner of their construction, and also of their fitting-up and furnishing, will exercise considerable influence upon their ventilation.

It has also been pointed out that, apart from definite air-inlets provided for ventilation, air may, and in almost every case will (particularly when open fires are in use), be drawn from sources frequently difficult to trace, and that such sources may be contaminated; and that, with the best possible arrangements provided for securing adequate change of air, efficient ventilation can only be secured by the constant attention and watchfulness of those who occupy the buildings, or of those entrusted with the care thereof, principally for the reason that change of air within a building is most frequently brought about by movements of the outer air, which constantly vary, and because comfortable change of air within is greatly affected by the temperature of that without, which also is subject to considerable variation.

Only by propelling air by mechanical means into a building is it possible to ascertain the source of the air-supply, to be able to cleanse it, temper it, and to regulate its hygrostatic condition, as well as cause it to circulate throughout the several apartments of a building without causing currents or draughts which may be unpleasant to occupants.

"**The Theory and Practice of Ventilation**", published in 1844 by David Boswell Reid, M.D., F.R.S.E., is replete with interesting facts and information, but partly, I am inclined to think, because he had not the appliances to work with which now are attainable, and partly because in some cases his observations have led him to wrong conclusions, his deductions are not to be fully relied on. This is most unfortunate, because I find so many recent writers on ventilation appear to take them all for gospel, and continue to disseminate views which now cannot be supported by facts.

The **general condemnation of downward ventilation** is a case in point; it has been called down-draught ventilation, and consequently condemned, for a down-draught is generally recognized to be objectionable, but it is now abundantly proved that, with air gently travelling in a downward direction, it can be supplied pure and with comfort even to each separate individual in a crowded hall; and, as previously explained, change of air in an ordinary apartment where there is an open fire, principally takes place in a downward direction.

The **placing of inlets and outlets on opposite sides of a room** is another of those errors, in past times originated, still handed down and followed by the unthinking. Windows are not intended to be included in this condemnation; as already explained, they may with advantage be on different sides of a room, as they can then be employed for quickly securing, at suitable times, a thorough change of air in the apartment.

In conclusion, let it be remembered that in ordinary dwelling-houses, whatever ventilating appliances may be considered advisable, it is essential that they be simple in construction, and easily regulated, and that anything in the form of ducts or flues must be of sufficient area to permit of ready access for cleansing, and must be periodically cleaned; and, as it is only by the employment of mechanical means that the flow of air in any one direction can be continuously regulated, and as such mechanical means cannot be economically employed in separate dwellings, there is at all times a liability from the varying temperature of the inner and outer air, or from the direction or force of the wind outside, to a reversal of currents, so that the inlets may become outlets and the outlets inlets; consequently, unless the outlets as well as the inlets are maintained in a clean condition, the air admitted may in its passage through them become contaminated, and so prevent the possibility of securing efficient ventilation.

SUPPLEMENTARY CHAPTER

BY THE EDITOR.

VENTILATION BY MEANS OF WARMED AIR.

Plates XX., XXI., and XXII., which illustrate two well-arranged houses erected from the designs of a London architect, Mr. E. J. May, have been introduced, with Mr. May's kind permission, for the purpose of showing what has been done in the way of designing houses with strict regard to their adequate ventilation. Such complete schemes of ventilation could not, of course, be adapted, without very great cost, to existing buildings, but in the case of new houses there is no reason why ventilation should not be considered from the outset, and its requirements be allowed to modify, to some extent, the design and construction.

Probably neither of these houses would have been erected as shown in the Plates, had not **Drs. Drysdale and Hayward** carried out somewhat similar arrangements in their own homes in Liverpool in the years 1861 and 1867 respectively, and had they not given publicity to their principles in papers read by Dr. Hayward before the Liverpool Architectural Society in 1868, and before the Royal Institute of British Architects in London in 1873, and in a book entitled *Health and Comfort in House-building*, which first saw the light in 1872. The principles advocated by these doctors have been already set forth herein in the Section on WARMING (pp. 104-7), and little need be added now. It must, however, be pointed out that they now disclaim all advocacy of warming houses by hot air; on pp. 27 and 28 of the *third edition* (1890) of their book, they make this clear: "We are no advocates of warming the house by means of heated air. All we recommend here is the warming of the incoming air that is requisite for ventilation. . . . The attention paid here to the distribution of warm air has led to the erroneous impression that our book recommends the plan of warming the house by hot air."

It will be seen, therefore, that the house at Chiswick, although directly receiving its initiative from the labours of the Liverpool doctors, went somewhat further, for it was **an attempt to combine warming and ventilation**,—that is to say, the air required for ventilation was to be warmed in the basement and distributed throughout the house as the sole warming agent; the kitchen-fire, which of course was required for cooking, supplied the necessary heat to the surrounding foul-air flue in order to create the up-current required to extract the vitiated air from the several apartments. This will be better understood by reference to the two Plates, and especially to the section and details given in Plate XXI. In the drawings no other fireplaces are shown, but Mr. May writes that fireplaces and flues were built, but the fireplace-openings were bricked up and therefore not used.

The fresh air was drawn into a special chamber in the basement, and there warmed by means of hot-water pipes slung to the ceiling of the chamber. Thence it passed through floor-gratings into the hall on the ground-floor, and the corridor on the first floor, and was then conveyed into the various apartments through holes in the walls, near the floors and ceilings.

The extract-flues were in proportion to the cubic capacities of the rooms from which they led; thus, the flue from the consulting-room was 10 inches by 5 inches, that from the kitchen 9 inches by 9 inches, and those from the two bed-rooms shown in the section on Plate XXI. were 11 inches by 6 inches and 9 inches by 5 inches respectively. The main extract-flue surrounding the smoke-flue had a clear area of 3·81 square feet, while the downcast-shaft leading to it had an area of 5·72 square feet.

As in Dr. Hayward's house, **the windows were "hermetically** closed", in order that no cold air might find direct access to the rooms, but that all the air required must pass through the warming chamber in the basement. In Dr. Drysdale's house—the earliest of the three—the windows were simply "close-fitting", but apparently the experience gained here led to the "hermetically-closed" windows of his friend Dr. Hayward's house, and later to those of Dr. Hogg's house at Chiswick, shown in Plates XX. and XXI.

There are several serious objections to such a combined system of warming and ventilation, especially when the windows are "hermetically closed" and fireplaces are conspicuous by their absence. In the first place, there is a want of elasticity about the arrangements; for example, it is often desirable, particularly in summer, to secure a large inflow of air to a room after it has been unduly occupied, as in the case of a dining-room where, after a dinner, the gentlemen have smoked abundantly; the best way of effecting the desired change of air is undoubtedly to open windows as wide as possible. In summer again, the currents of fresh air through the house, caused by open windows and doors, are often positively refreshing. In such a house as that at Chiswick, no refreshing changes of air could be rapidly accomplished. Then again, at night,

when ventilation and fresh air are even more necessary than during the day, the out-doors will be closed, so that no air can find entrance except through the warming chamber, and the kitchen-fire will not be burning, so that the current in the extract-shaft will be often sluggish in the extreme, and the supply of fresh air to the rooms and the extraction of vitiated air will be imperfect. Air-ducts, moreover, may become coated with dust and organic matter, and may actually pollute the air passing through them, unless they are so arranged as to admit of ready cleansing, and unless the cleansing is regularly done.

More might be said, but perhaps a quotation from Mr. May's letter, with reference to the house at Chiswick, will suffice: "A family lived in it safely for four years, but I expect they found it dull comfort to have nothing but hot-air gratings to look at, and no usual draughts to grumble at, so at the end of that time the windows were made to open, and the fireplaces (which I had had formed) opened out and used, thus bringing this house practically to the same scheme as the Hampstead one".

The **Hampstead house is shown in Plate XXII.**, and undoubtedly avoids the principal defects of the earlier building. The drawing illustrates a pair of houses, but only the right-hand house includes the special arrangements for ventilation. The arrangements for the warming of the incoming air and its admission to the several corridors and apartments, and for the extraction of the vitiated air by means of upcast flues leading to a foul-air chamber in the attic, and thence by a main down-cast shaft to the basement, and up again by an extract-shaft surrounding the kitchen flue, are practically the same as those at Chiswick, but in addition to them, the windows are made to open, and fireplaces are provided as usual in the rooms and used when required.

This scheme appears to have worked satisfactorily: at any rate, Mr. May is able to write: "My client always said he derived great benefit and comfort from the system". Undoubtedly it possesses merits which were wanting in the other house, notably the merit of elasticity. If, in case of sickness, the temperature of any room requires raising, it can be done by means of the open fire provided in that room; increased ventilation can at any time be quickly provided by opening the windows, and additional warmth combined with greater extraction of air by lighting a fire in the grate.

Somewhat similar houses have been erected in various parts of the country, but it is certainly somewhat surprising that, after the lapse of more than 30 years from the erection of Dr. Drysdale's house in Liverpool, so few examples occur in which the principles have been carried out. Either houses as ordinarily built are not so bad as they are painted, or the proposed system has serious practical defects, or architects and householders are grievously slow to learn. Perhaps all three reasons have something to do with the tardy adoption of the system. Mr. May, although he has evidently studied the system closely, and has had practical experience in the working of it, is not by any means enthusiastic about it; he says: "I have never professionally advocated it, nor, since I built these houses, have I designed any others like them, nor do I expect to very much".

Another arrangement for supplying warm air to the several rooms of a building is shown in Plate XXIII. It has been designed by Mr. Wm. Bruce, whose gas-stoves or radiators have already been mentioned. In this larger apparatus also, gas is used as the heating medium, more or less "sections" of the apparatus being used according to the work to be done. The illustration shows ten of these sections enclosed in a chamber of glazed brickwork. Access-doors are provided. A separate supply of air is conveyed to the burners by means of the pipes marked M, and a special flue J is constructed for carrying off the products of combustion.

The air to be warmed is brought from the exterior through the large duct C, in which are placed three filtering screens D. A fan is sometimes inserted at E for propelling the air, and when this is in use the flap F is lowered to the position shown by the dotted lines. Between the fan and the warming chamber, water, mixed with disinfectants if desired, can be sprayed through the incoming air, so as to purify it still more; this is effected by means of the two cylinders marked G on the end elevation, and the adjacent pipes. After passing through the warming chamber, the air is led along the ducts K K K to the various apartments.

This apparatus provides means for purifying the air, and has other advantages, which should render it of considerable utility for ventilation, provided that suitable extract-shafts are provided, and that the air is properly humidified after being warmed.

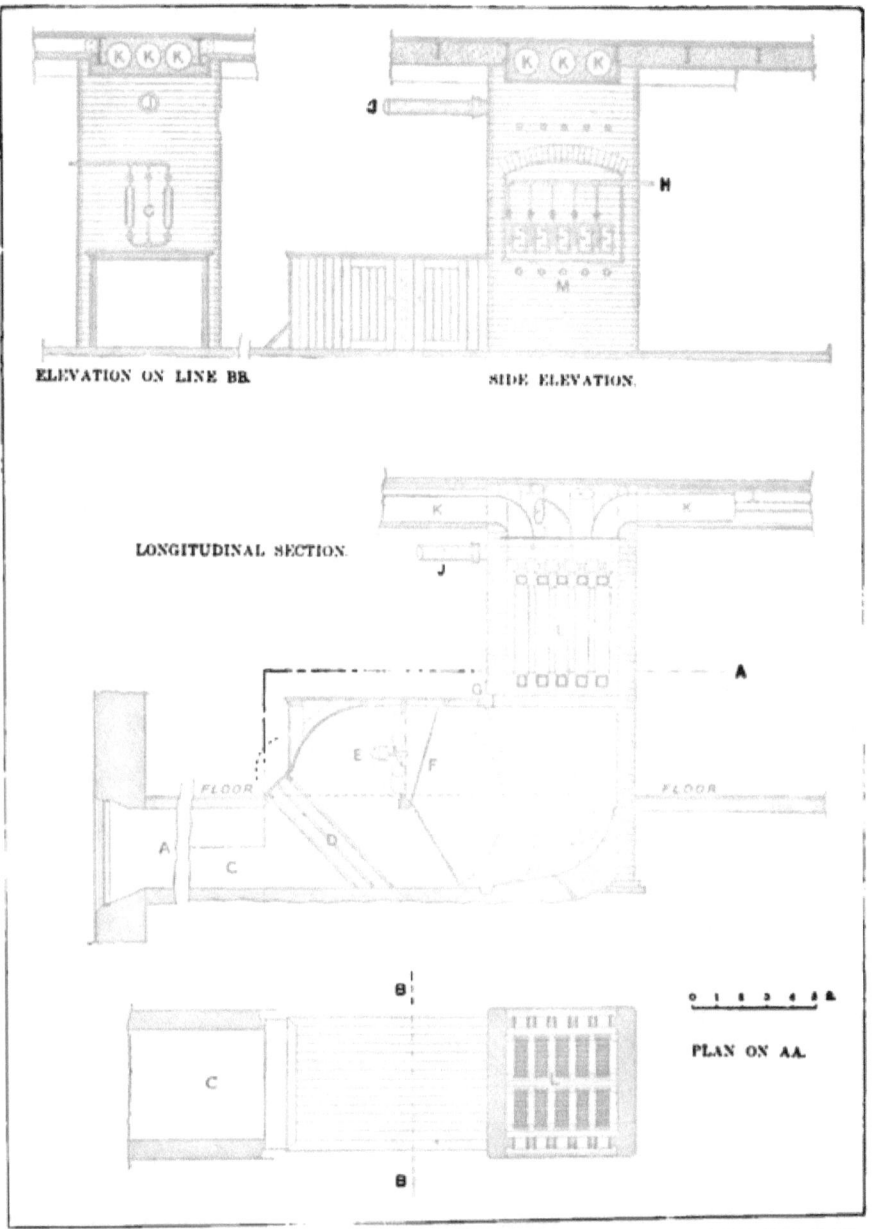

BRUCE'S WARMING AND VENTILATING APPARATUS, HEATED BY GAS.

C. Main air-duct.
D. Screens.
E. Fan.
F. Flap.
G. Humidifying and disinfecting apparatus.
H. Gas-inlets.
J. Flue for waste products.
K. Ducts for heated air.
L. Ten "sections" heated by gas.
M. Air-inlets to gas-burners.

Section XIII.—Lighting

Part I.—Candles, Oils, and Electricity

BY

E. A. CLAREMONT

MEMBER OF THE INSTITUTION OF ELECTRICAL ENGINEERS, MEMBER OF THE INSTITUTION OF MECHANICAL ENGINEERS
AUTHOR OF "ELECTRIC LIGHTING", ETC.

Part II.—Gas

BY

HENRY CLAY

FIRST HONOURS IN PLUMBING, CITY AND GUILDS OF LONDON INSTITUTE, REGISTERED INSTRUCTOR IN PLUMBING
AUTHOR OF "PRACTICAL PLUMBING", "HOT WATER FITTING", ETC.

SECTION XIII.—LIGHTING.

PART I.—CANDLES, OILS, AND ELECTRICITY,

BY

E. A. CLAREMONT, M.I.E.E., M.I.M.E.

CHAPTER I.

CANDLES AND OILS.

The illuminating power of candles is subject to great variations, owing partly to their mode of construction, and also to the great irregularities caused by the lengthening and shortening of the wick; thus, when a freshly-snuffed candle is first lit, there is a rapid increase in the power of the light, and this increase continues until the burnt wick leaves a residue of waste and spongy matter, from which all combustible material has been extracted. This, if not removed by snuffing, will cause the light to decrease as steadily as at first it increased. Such variations in the burning intensity of candles have been found by experiment to reach the following almost incredible ratios:— A candle which, when first lit, gave an intensity of 100, gave in four minutes about 90, and in twenty minutes only 30. It can be seen from this that such an illuminant cannot be handled, for the sake of comparisons and experiments, with the same facility with which we can treat lights of a more regular character. Hence, beyond its use as a mere verbal standard of power, we may dismiss this form of lighting altogether.

In England, however, as the expression **"candle-power" (C.P.)** has reached a somewhat similar position in the comparison of intensities of light to that possessed by Carsel's lamp in France, I may as well state here that this standard was originally derived from a spermaceti candle, consuming 120 grains per hour, and that all relative tests are made, not directly by measuring the relative intensities of the flames, but by gauging the relative depths of the shadows thrown by them.

Mineral burning oils, usually purchased under such names as "Daylight", "Sunlight", "Kerosene", &c., are formed from the differently-refined products of petroleum and shale oils. By "petroleums" are usually meant such oils as are obtained from springs, in their crude or liquid form, while the shale products are those distilled from oil-producing turf and rocks; although the word "petroleum" is usually applied to the former, and "paraffin" to the latter, we may really look upon these names as mere British trade-distinctions. The word "kerosene", which was formerly used to distinguish a particular brand of Russian oil, is now applied to mineral burning oils generally, at least to such of them as are suitable for burning with wicks. Kerosene is a colourless liquid, insoluble in water, lighter than water (upon which it will therefore float), and usually possesses a well-defined blue fluorescence. Good kerosene may also attain a yellowish tint, but this should not be too pronounced, as it would point to the presence of heavier oils, which, to be consumed satisfactorily, would require a wick of looser texture than that employed for the finer samples. It should be noticed, however, that the yellow appearance may be caused simply by undue exposure to light.

In England, all Government requirements, regulating the storage, transportation, &c., of burning oils, are regulated by what is known as their "**flashing-point**". This is simply the degree of heat at which the oil begins to give off sensible quantities of inflammable vapour. This method of deciding the inflammability of oil is far superior to the method still much in vogue in America, where the "burning-point", or temperature at which an oil ignites bodily, is considered sufficiently reliable for ordinary purposes. As, however, the burning-point depends very largely on surrounding circumstances,—such as the pouring of the oil over a broad porous surface, when it will ignite much more easily,—such a standard gives little idea of the nature of the oil.

In the case of cold kerosene, neither the oil nor its vapour ignites on the application of a light. In the case of lamp-explosions, such vapour must have been given off at a higher degree of temperature than the flashing-point of the oil (probably through heat conducted by the body of the lamp), when, of course, such vapour becomes explosive. In England, as the result of numerous comparative experiments made with two forms of apparatus, of which that known as the "close test" is the one usually employed, the flashing-point of "petroleums", as kerosenes are termed in the statute-book, has been fixed by law at 73° Fahrenheit. This standard, though considered by some to be too fine owing to the best of the American oils being kept for home consumption, does not really affect the better qualities of the kerosene imported into this country, of which the flash-point is not usually less than 100° to 120° Fahrenheit.

Russian oils give a steadier light than American, as they do not seem to possess the same mixture of light and heavy constituents, which usually cause the American oils to burn at first with their greatest brilliancy, gradually diminishing as the oil in the reservoir decreases and the heavier portion is reached. Russian oils, however, though steadier, never give as much light.

Owing to such variations in the density of kerosenes, it will be seen that great care should be exercised in **the choice of wicks**, for, especially in the case of the heavy oils, a wick of loose and pliable texture, such as is generally made from long-staple American cotton, is infinitely preferable to others of a harsher nature.

In the consideration of oil-burning as an illuminant, we enter upon **the most dangerous phase of lighting.** From the most recent experiments by experts, it would seem that, from a representative collection of ordinary oil-lamps purchased in England and subjected to rigorous tests, those considered as dangerous or unsafe reached the extraordinary total of 90 per cent of those bought. From the list of casualties published by the Public Control Department of the London County Council, we see that in 1894 the fire-brigades were called to no less than 448 fires in the metropolis alone, all of which were ascribed to oil-lamp accidents. Of this number 337 were caused by the upsetting of lamps, while only 90 were brought about by explosions. Thus it seems that, in the use of oil-lamps, the danger most to be feared is that occurring through the leakage of oil from the reservoir after upsetting, and this we find from the most recent statistics is caused by lamps being built of fragile material, burners being attached by screws whose threads are of an imperfect nature (or on which there are no threads at all), and by sconces holding wicks of imperfect fit, thus allowing an outlet through which the oil may flow. On the other hand, explosions, which would seem to be of much less frequent occurrence, are chiefly caused through the vapour in the reservoir becoming inflammable, on account of the temperature of the oil having exceeded its flashing-point, this being generally due to heat being conducted through the body of the lamp from the burner. It would seem, therefore, that the material of which the reservoirs are composed should be of non-heating capacity, but as this would result in their being constructed either of glass or porcelain, it would lead to a worse evil, namely, having an apparatus of a fragile nature.

A good lamp will fulfil the following conditions:—

1. It will be made of a material not easily broken.

2. Such material, if of metal or other heat-conducting substance, will be insulated from the burner by the use of as long a neck as possible.

3. The actual burner will be continued in the form of a tube containing the wick to within a very short distance from the bottom of the reservoir, thus preventing both the falling of a hot wick into the oil, and the ignition of vapour through contact with it.

4. The reservoir will be made to clip into, and not be loosely borne by, the sconce or cup into which it fits.

5. If the lamp be a high one, its stand or pedestal will be heavily weighted with lead at the bottom.

6. The burner will be screwed on, and not merely cemented or riveted, and such screw will be composed of a good thread for at least four complete turns round the neck of the reservoir.

In the actual use of the lamp, great care should be taken that only **a tightly-fitting wick** is used; especially is this the case with round burners, into which only endless wicks, or those perfectly cylindrical in shape, should be fitted in order to avoid the escape of oil previously referred to.

From careful **comparison with other illuminants**, it would seem that oil takes a high place on account of its comparatively slight contamination of the surrounding atmosphere by carbonic acid and other gases. Its heating effects are equally low, though of course in brilliancy it has been far superseded both by electricity and the most modern forms of gas-burners. The approximate cost of using a typical oil-lamp is somewhat similar to that of an Argand gas-burner, as shown by the following table, in which oil, gas, and electricity are compared:—

TABLE XXXVIII.

COMPARISON OF THE COST OF ILLUMINANTS.[1]

Illuminant	Price	Burner	Candle-power	Amount of Illuminant consumed per hour	Cost per C.P. per hour, in pence
Oil,	8d. per gallon = 7 lbs. 11 ozs.	Lamp	16	2·08 ozs.	·0084
Gas,	2s. 6d. per 1000 cub. ft.	Ordinary Bray gas-burner	9·6	6 cub. ft.	·0187
,,	,,	Argand burner	15·8	4·5 cub. ft.	·0085
,,	,,	Welsbach incandescent burner	20	4 cub. ft.	·006
Electricity,	2d. per unit	Electric incandescent lamp	16	60 watts	·0075
,,	4d. ,, ,,	do.	16	do.	·015

[1] The following table, translated for this work from the *Handbuch der Praktischen Gewerbehygiene* (Berlin, 1896), but slightly altered in arrangement, goes more fully into the comparison of illuminants, including not only

Bracket, pendant, and table oil-lamps, especially the more expensive kinds, generally have their reservoirs resting in cup-shaped sconces of more or less elaborate workmanship, and fitting therein very loosely to allow of their ready removal; this constitutes a source of considerable danger. While, however, the brackets, pendants, &c., vary greatly in form and construction, the types of lamps used for them vary very slightly. **Hand-lamps** are usually safer, as the handles are almost universally attached direct to the reservoirs, thus adding greatly to their stability, and as the reservoirs are usually constructed with much broader bases than other kinds of lamps.

In conclusion may be noticed the two most common **methods of suspending and elevating lamps**, the apparatus being fitted either direct to the lamps, or (more frequently) to the cup-shaped holders above described. The method most frequently in use is constructed somewhat in the form of a pair of scales, having a long lever suspended by a chain from the ceiling in such a manner that a lamp fixed to one end balances a weight on the other; when the lamp is pulled down the balance-weight goes up, maintaining that position until the lamp is again moved. This apparatus has also a circular horizontal movement, allowing the lamp attached to be swung round in a circle of about 6 feet. The other apparatus consists of a small metal case hung to the ceiling, and containing a thin steel ribbon, which is acted on by a spring adjusted to a tension corresponding to the weight of the lamp, by means of a thumb screw; the lamp,

the question of cost but also the important question of the products of combustion. It will be noticed that the round-wicked oil-lamps are, according to the figures in the table, the most economical form of lighting, while the Regenerative and Argand gas-burners are both cheaper than electric glow-lamps. The Welsbach incandescent gas-burner does not appear to have been tested; it would undoubtedly have shown more economical results than any of the other burners mentioned. Of course, the products of combustion must be considered as well as the cost, and it is in this respect that electricity possesses the greatest merit, and that the ordinary batswing gas-burner shows to the greatest disadvantage.—ED.

TABLE XXXVIIIA.
GENERAL COMPARISON OF ILLUMINANTS.

Illuminant.	How burnt.	Consumption to obtain a light of 100 candle-powers per hour.	Cost per hour in pence.	Products of combustion for equal amounts of light.		
				Water in Kg	CO_2 in cb. metres	Heat-units
Electricity,	Arc lamp.	·09 – ·25 horse-power	·6 – 1·44	...	...	100
"	Incandescent lamp.	·46 – ·85 "	1·80	...	...	300
Gas,	Siemens' Regenerative burner,	·35 – ·56 cub. metres	·72 – 1·20	...	...	1,500
"	Argand burner,	·8 – 2·0 "	1·74	·66	·46	4,680
"	Batswing burner.	2·0 – 5·0 "	4·32	2·14	1·14	12,150
Petroleum,	Round or "solar" wick,	·25 kilog.	·60	·27	·45	3,360
"	Flat wick.	·60 "	1·29	·80	·95	7,200
Refined Oil,	Round wick.	·25 "	·54	·37	·45	3,360
"	Flat wick.	·60 "	1·27	·80	·90	7,200
Rape-seed Oil,		·50 "	4·98	·52	·71	4,300
Stearine,	Candle.	·90 "	19·72	1·04	1·20	9,000

being fastened to the end of the ribbon, remains suspended at any distance to which it may be pulled from the ceiling.

CHAPTER II.

ELECTRICITY: DEFINITIONS.

At the outset it is necessary to define a few of the technical terms in use among electricians.

Volts, Voltage, Electro-motive Force, Potential,—all signify pressure; just as water is delivered at so many "lbs." pressure, electricity is delivered at so many "volts" pressure.

Ampère and Current both signify quantity; as in speaking of water, one would say so many gallons, so in speaking of electricity, one would say so many ampères, and to combine the expressions of pressure and quantity, one would say, for instance, 10 ampères at 100 volts.

Watts are the product found by multiplying ampères and volts together; for example, 100 volts multiplied by 10 ampères represent 1000 watts, which is a certain amount of **electrical energy**. If a battery were described as giving 1000 watts at (say) 50 volts, the current could be ascertained by dividing the 1000 by 50, the result in ampères being 20. So any quantity of energy, described by so many watts, can be expressed in volts and ampères, provided the quantity of either of these latter is known.

Ohm is the term used to express the resistance offered in any electrical circuit. This resistance—to use again the analogy of water—may be compared to the resistance offered by a pipe to a quantity of water passing through it.

In order to illustrate the meaning of these terms, we will take one example comprising them all, using the analogy of water again to prevent any misunderstanding. If 100 gallons of water at a pressure of 50 lbs. are passing through a pipe offering a certain resistance, the 100 gallons multiplied by the 50 lbs. give a force of 5000 pound-gallons, or a definite amount of energy, from which, however, the resistance offered by the pipe remains to be taken; so, with electricity, we can have 100 ampères at a pressure of 50 volts, passing through a certain resistance, and the 100 ampères multiplied by the 50 volts show a force of 5000 watts, or a definite amount of energy, while the positive side of the resistance will be that at which the current enters, and the negative side that at which it leaves, or to which the energy of the current has been reduced or exhausted.

All calculations, such as those above given, follow a rule known as **Ohm's law**, which is a rule for ascertaining any one of the three quantities, if the other two are known. Let C = current, E = electro-motive force (usually abbreviated to E.M.F.), and R = resistance, then

$$C = \frac{E}{R}, \quad R = \frac{E}{C}, \quad \text{and } E = R \times C.$$

As an example, we will suppose that a current at an E.M.F. of 50 volts is passing through a resistance of 5 ohms, and that it is desired to know the quantity of that current; this is ascertained by merely dividing the 50 volts by the 5 ohms, which gives the result of 10 ampères. Similarly, the resistance or the electro-motive force can be ascertained if, in each case, the other two quantities are known.

The terms "**positive**" and "**negative**" refer to the direction of the current, the former being often designated by the sign +, and the latter by the sign −.

CHAPTER III.

ELECTRICITY: GENERATION AND STORAGE.

The simplest method of obtaining current in practical form for electric lighting is, no doubt, **the primary battery**, but owing to the small amount of energy obtainable from a battery of reasonable dimensions, the unpleasant process of cleaning and renewing, and the somewhat great expense of maintenance, I do not purpose to consider it in detail. Many forms of batteries have been put upon the market, especially during the last few years, some of them showing very great ingenuity, but all with which I have come in contact have at least one of the above-mentioned disadvantages to a sufficiently marked extent to condemn them for electric lighting on an extensive scale.

The dynamo, which has made such rapid strides towards perfection since it became recognized as a commercial article about eighteen years ago, is the means now almost invariably adopted for obtaining electric current. It can be best explained by reference to a steel magnet, which is merely a piece of tempered steel magnetized (originally by a lodestone), and is generally either straight or of horseshoe pattern. Between the poles of a magnet there are always a number of "lines of force", which the accompanying illustration will make clearer than any description. This illustration was obtained by placing the small magnet shown, upon a table, and then covering it with a piece of cardboard having upon

Fig 610 Magnet and Lines of Force.

its surface a quantity of iron filings, which immediately placed themselves as seen. Any magnetical metal coming within the range of these lines of force will be attracted to the matter, and will itself carry some of the lines and become a magnet. Any non-magnetical metal will, of course, be quite unaffected as far as magnetism is concerned.

But supposing we inserted and rotated in the lines of force between the poles of a magnet, a ring of copper (such as a key ring), there would be induced in the copper ring an electric current of minute quantity. This phenomenon of induction would be produced without any friction, or any treatment beyond its rotation within the limits of the lines of force. It will at once be seen that, by having a powerful magnet and many large coils of wire, a powerful current could be produced, and this fact has been taken advantage of in the dynamo, which is nothing more than a magnet, with coils revolving in the lines of force created.

A dynamo, however, has not a magnet such as that just described (which is known as a permanent magnet), but has what is known as **an electro-magnet**

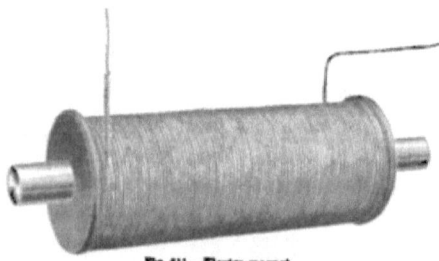

Fig 611.—Electro-magnet.

—shown in fig. 611,—as this can be produced more strongly and cheaply than a permanent magnet, and offers other advantages which will be seen hereafter. There is, however, no difference between a permanent magnet and an electro-magnet, as far as the lines of force are concerned, and so the principle remains the same. An electro-magnet differs from a permanent magnet in that it is not made of hard steel, but of soft iron,—and sometimes soft steel,—and has wound upon it many turns of insulated copper wire, through which a current is passed, which renders the iron, *as long as the current passes*, strongly magnetic.

The coils which revolve in the lines of force of such a magnet are termed **the armature**. This armature generally consists of a number of discs of thin

THE DYNAMO.

charcoal iron,—each with a hole in its centre concentric with its circumference,—which are slipped upon a steel shaft, and upon the periphery formed by

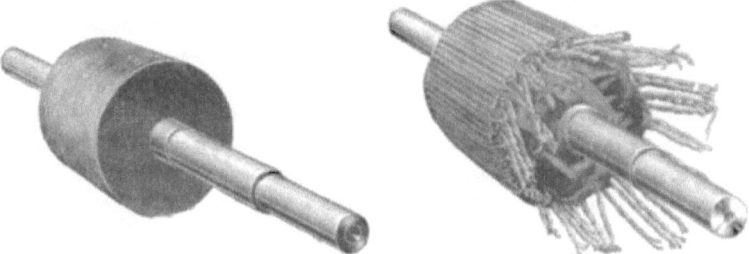

Fig. 612.—Shaft, with Discs in Position, forming Armature.   Fig. 613.—Armature Partially Wound.

these plates are laid the insulated copper wires forming the coils, which are to revolve between the poles of the magnet.

These coils are connected to a number of copper plates on edge, radiating from the shaft, as shown in fig. 614, and separated from one another by insulation of some sort, preferably mica. The copper plates form what is known as **the commutator.** Upon this commutator, at diametrically opposite points, brushes are brought to bear, which collect the current as it is generated in the coils revolving between the pole-pieces.

Fig. 614.—Armature and Commutator.

As I have mentioned, **the electromagnets of a dynamo** require a current to be passed round them, to enable them to give the required lines of force. This current is obtained from the armature, and is very small when compared with that which the armature is capable of giving out.

A question often arises in the student's mind as to how it is possible for the dynamo to commence work, seeing that the magnetism of the poles is dependent on the current in the armature, and the current in the armature on the magnetism of the poles; for, as above described, to become magnetic, iron must have a current passed round it. How then is the dynamo to start itself?

The explanation is that, no matter how soft the iron of which the magnets are made, they generally retain sufficient residual magnetism to cause them, to a slight extent, to act as permanent magnets, and (without the winding upon them having a current passed through it) to produce a sufficient number of lines of force to create a small current in the armature, which then passes through the winding upon the magnets, and strengthens them sufficiently to induce more current in the armature; thus the machine on being "started up",—that is to say, driven by an engine,—gradually "builds up". With dynamos having a large quantity of iron in their magnets, the time required to magnetize the poles and "build up" the machine is very appreciable.

Dynamo-builders occasionally produce a machine, which, on being driven by an engine for the first time (to test it), refuses to show the slightest magnetism, and it is then necessary to magnetize it from another source, perhaps the current from a battery, or from another dynamo which may be at work and producing current.

There are three kinds of dynamos—"*Series*", "*Shunt*", and "*Compound*", which are all built in various patterns. It will suffice to describe simple diagrams of the three kinds, as the theory, which will be readily followed, can be applied to any machine, no matter who the builder or what the pattern.

1. **The "Series" dynamo** is so called in consequence of the current from the armature passing *in series* round the magnets and outer circuit. By the outer circuit is meant the path which the current traverses, after leaving the machine. Fig. 615 shows the commutator in section with the brushes bearing thereupon. By following the path from the upper brush, it will be seen that the current, on leaving the armature, passes through the wire wound on the limbs of the magnet, and then proceeds through the external circuit and back to the other brush.

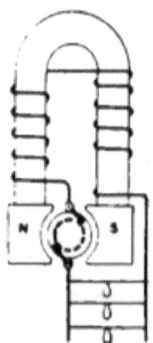

Fig. 615 — Diagram showing Series Dynamo and Circuit.

In this particular diagram there are three paths in the outer circuit through which the current can pass, and the quantity taken by any path will depend upon its resistance compared with the resistance of the others. Naturally, the lower the resistance of any path the greater will be the current through it. With a constant pressure and known resistance of any path, the current through it can be calculated by Ohm's law, as already described.

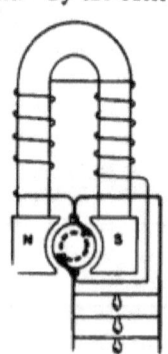

Fig. 616 — Diagram showing Shunt Dynamo and Circuit.

2. **A "Shunt" dynamo** is so called in consequence of the circuit of the magnets offering a "shunt" path to the external circuit. As will be seen from the diagram (fig. 616), the current leaving by one brush can go direct to the

external circuit, and return therefrom to the other brush without having to go through the magnet-winding, as in the series machine. The shunt or magnet-winding also offers, similarly to the external circuit, a complete path, and so the current also passes through it. Since, then, there are two paths for the current from the armature—viz. the external circuit and the magnetic-winding,—the current through each will depend upon their relative resistances, and the shunt-winding, being made of higher resistance, takes only a little of the energy.

In a *series* dynamo, all the current that passes from the armature goes around the magnet, and **the wire forming the magnet-winding** must therefore be of a larger section than that on a shunt machine, which only requires a small quantity of current to pass around the magnet. For the purpose of explaining this point, we will suppose that we have a series and a shunt dynamo, each of which has an armature with wire upon it of sufficient diameter to give 100 ampères with a certain excitation of the magnets. Each machine requires a certain amount of excitation, and supposing, in the case of the series machine, that, by providing 50 turns of wire on the magnets, the 100 ampères which must pass through them produces the amount of excitation in question, that will be 50 turns multiplied by 100 ampères, or 5000 ampère-turns. Now a similar number of ampère-turns on the shunt machine will give in its case the excitation required, but we cannot pass 100 ampères (the total current from the machine) through it. So we take a much finer gauge of wire, which will, with (say) 2000 turns, offer such a resistance as to allow only $2\frac{1}{2}$ ampères to pass through it, when it will be found we have a similar excitation to that of the series machine, since 2000 turns multiplied by $2\frac{1}{2}$ ampères = 5000 ampère-turns, or a similar amount of exciting energy.

I mentioned above that the armature of each machine gave 100 ampères, and should like to make it clear that **the limit of current which a wire will carry**, whether on an armature or elsewhere, is that amount which will create in the wire just such sufficient heat as *not* to damage its insulation. The insulation may eventually suffer without the heat at any particular time appearing excessive, but upon this point I shall be able to say more later.

Comparing the two types of dynamo described, "Series" and "Shunt", we find the former has this peculiarity: that the more the paths offered to the current in the external circuit, or, in other words, the less the resistance of the external circuit, the greater will be the quantity of current which will pass through that external circuit, and consequently through the magnet-winding. But the greater the quantity of current through the magnet-winding, the greater will be the number of lines of force created to act upon the armature; so the less the external resistance, the stronger the machine. To see this plainly, we have only to turn to the diagram of the "series" dynamo, and cover up the external circuit with a piece of paper, and imagine a series machine without any circuit, when it will be obvious that no current can pass through the magnet-winding at all, since there is no path for it back to the armature; and it will also be obvious that, if we only provide an external circuit of very high resistance, only a little current can pass through it and consequently around the magnet.

Now in the case of the "shunt" dynamo, unlike the "series", the less the external resistance the weaker will be the machine, since the amount of current flowing in the magnet-winding and the external circuit depends upon their relative resistances. To see this plainly, if, as in the case of the diagram of the "series" machine, we cover up the external circuit of the shunt dynamo, and imagine it without any, we shall see that the magnet-circuit alone offers a path back to the armature, and will consequently take all the current it requires; but, on our providing another path, the current in the magnet-circuit will be decreased; hence, the machine is weakened as the external circuit decreases in resistance.

Turning again to the diagram of the series dynamo, if we were to join the brushes across with a piece of copper wire, we should be said to have "**short-circuited**" **the armature**, since the current would take a shorter path than that originally provided for it. Now let us see what happens if we short-circuit the terminals, *i.e.* the two points at which the external circuit commences. We have just seen that the lower the resistance of the external circuit, the stronger the machine. Here is a case where the external resistance is practically nothing, since we have short-circuited it; the machine, therefore, will generate current to excite the magnets, which will in turn produce more current, and the amount of current produced will, under such conditions, soon burn up the insulation of the armature-winding.

If we short-circuit the terminals of a shunt machine,—that is to say, if we join the two wires across as they leave the dynamo on the external circuit,—the result is very different, since the machine will no longer give any current at all. The reason being that we have also short-circuited the magnet-winding, since both magnet-winding and external circuit emanate from the same points, the brushes; whereas, in the series machine, the current has to pass through the magnet-winding before it meets the external-circuit wires.

3. **The Compound Dynamo.** In electric lighting, it would be very inconvenient to have a dynamo which, when driven by an engine running at a

constant speed, varied its power and pressure according to the quantity of light burning, and yet this is what happens with "series" and "shunt" machines, since the former becomes stronger, and the latter weaker, on an increase of current being required in the external circuit. In order to avoid this inconvenience, the compound-wound machine has been designed, which, as its name implies, is a combination of the "series" and "shunt" dynamos, the magnets being wound in both ways, and in such a proportion, that, although the resistance of the external circuit varies, the pressure of the machine remains constant. It will be understood from the diagram of this compound machine, fig. 617, that, on the external circuit decreasing in resistance through more lamps, &c., being turned on, the shunt-winding weakens the magnets, while that of the series strengthens them, and so a balance is kept.

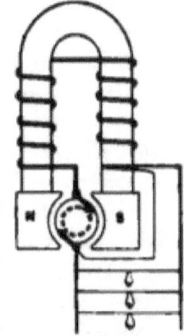

Fig. 617. Diagram showing Compound Dynamo and Circuit

If, with a shunt machine, constant pressure should be necessary, it could be obtained by regulating the resistance of the magnet-winding to suit the resistance of the external circuit. For instance, supposing the external circuit were to decrease in resistance as we have seen, less current would traverse the magnets, and so the machine would become weaker. But if the resistance of the magnet-winding could be decreased also, the amount of current round the magnets would be maintained. To effect this adjustment of the magnetic winding on a machine giving 100 volts at (say) 1200 revolutions per minute, the magnet-winding should be cut and **an adjustable resistance,** such as is shown in fig. 618, inserted; that is to say, the current traversing the magnet-winding would have to traverse the resistance-frame before proceeding through the remainder of the winding.

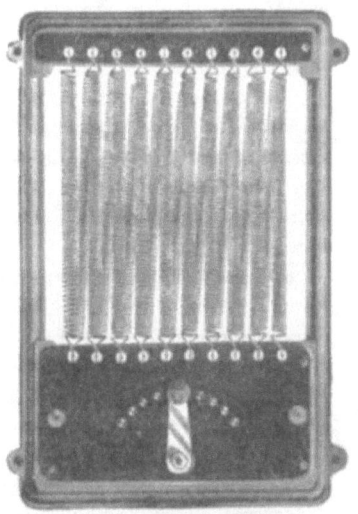

Fig. 618.—Shunt Resistance-frame.

The resistance in circuit with this winding would depend upon the position of the movable switch-lever, which is so arranged that, by moving it over the different studs on the frame, any number of resistance-springs can be thrown in or taken out at will. If the switch-lever is drawn over to the left as far as the first stud, the whole resistance is short-circuited, since the current enters at the left-hand terminal, and passes, by way of the first left-hand stud, through the switch-lever and the right-hand terminal, without having been through any of the resistance-springs. If, however, the lever is drawn on to the right-hand stud, the current will have to traverse all the resistance-springs before arriving at the left-hand terminal.

With such a dynamo and resistance-frame, if the machine is driven at (say) an extra 50 revolutions per minute, making 1250, at which perhaps it would give 105 volts, sufficient resistance should be inserted to reduce this voltage to 100 as originally required. Again, on the voltage dropping through the external resistance decreasing, a little resistance could be cut out to bring it up to 100 again; and similarly, on the voltage increasing through lamps being turned off, resistance could be inserted.

These resistances are frequently used in the shunt-winding of compound dynamos, as well as in plain shunt machines, because, although a compound machine should regulate itself to within about two volts of full load on "open circuit" *at a constant speed*, yet, in many instances, the speed of the engine cannot be "governed" thus constantly.

A resistance-frame is frequently fitted with an electric device, by which it automatically adjusts the potential of the dynamo; when the E.M.F. drops, the apparatus moves the lever, or its equivalent, in such a way as to automatically short-circuit certain of the resistance-springs, thereby bringing the pressure up again. Such an appliance is called an automatic dynamo-regulator or resistance.

Of the three types of dynamos mentioned, *series* machines are little used except for the lighting of arc lamps, themselves in series,—that is to say, the current from one terminal of the dynamo traverses every lamp in succession, and then returns to the other terminal. For transmission of power, and for charging accumulators, *shunt* dynamos are principally used; also for the generation of current in electric-light stations or works, as they readily allow of being joined in "parallel", that is to say, two or more such machines can be connected to one external circuit, by connecting all their terminals on one side to one wire of the external circuit, and all their terminals on the other side to the other wire. The dynamos would require, in such an arrangement, to be all of similar potential or voltage, and supposing the potential of each were 100, the total potential would be 100. The current, however, from the combined machines would be the sum of all the currents; if there are five machines, each generating 100 ampères, the total will be 500 ampères.

I have throughout spoken of the current leaving one terminal of a dynamo, flowing through the external circuit, and returning through the other, and I may here point out that the terminal at which it leaves is called "*positive*", and that at which it returns is called "*negative*", the former being often designated by the sign +, and the latter by the sign —. The question as to which terminal will be positive, on a dynamo being taken to the testing-room to be run for the first time, is decided by the way its ironwork has become magnetized, and so, unless precautions are taken, it is a mere chance which is positive; should the right limb of the magnets become the "North" pole, and the left the "South" pole, the machine could be "reversed", as it is termed, by reversing the magnetism of the poles. This could be done by reversing the current through the shunt-winding of the magnets.

In speaking above of shunt dynamos being connected in parallel, it should be understood that the "positive" of each machine would constitute the one side mentioned, and the "negative" the other side.

Since, as I have shown, **the polarity of a dynamo** can be reversed by reversing the current through the magnet-winding, care must be taken to prevent the polarity of one of a number of machines working in *parallel* becoming reversed accidentally, because in such an event it would not be like the remainder, sending current in one common direction, but would be acting in just the contrary manner; the reversed machine would be absorbing current given out by the others for the circuit. Fortunately a shunt machine cannot be so reversed, since, if the positive brushes of two machines are joined together, the shunt-windings of their magnets are both receiving positive current, and if one machine stopped, the other would still be supplying positive current at that point into its magnet-winding. Whereas, with series machines, if one machine stopped, the current from the other would immediately traverse its magnet-coils in the opposite direction to that in which the current from its own armature had, a moment before, been passing, and the magnets would become reversed, necessitating the reconnecting of the terminals. It is for this reason of immunity from reversing, that shunt dynamos are used in works for generating electricity.

Compound machines, being partly series-wound, can not be joined in parallel without precautions being taken too complicated to mention here.

Either series, shunt, or compound **dynamos can be joined "in series"**, which means that the current from one machine is made to traverse the other before arriving at the external circuit. As I mentioned just now, machines in parallel give a total E.M.F. of any one of them, and a current of all their currents added together. Machines in series, however, give a current of any one of them, and a total E.M.F. of all their E.M.F.'s added together.

Machines in parallel must all be of the same E.M.F. to prevent one "running back" on another; they can, however, give varying quantities of current. Machines in series must all give the same current; otherwise one would be subjected to a greater current than that for which it was built, since the current from each machine passes through the remainder.

Compound dynamos are chiefly used for **private electric-light installations**, having, as I have described, the important feature of maintaining, with a constant speed, a constant E.M.F. The value of a constant pressure is recognized by all, since the fluctuations in the light given by incandescent lamps is chiefly due to slight variations.

As I have mentioned **transmission of power** above, and as electric power is now being more and more largely used, not only in workshops and factories, but also in houses and vehicles, a few words may be added to describe its theory. If a dynamo be driven by an engine, and give (say) 100 volts, and two wires therefrom be taken to a similar machine (called the "motor"), and placed in its terminals, the motor will revolve its armature at such a speed as to give a counter E.M.F. of nearly equal quantity, and will give off at its pulley a power nearly equal to that expended upon the dynamo.

Power-transmission on this principle has been largely adopted of late. In many instances where manufacturers have boilers supplying a large number of engines throughout their works, it has been proved most conclusively that, by substituting one main engine driving a dynamo, by which to supply with current motors in the place of the engines previously used, great economy is effected, owing to the avoidance of loss by radiation, and the absence of small steam-engines, which are notoriously not economical. The principle is of course without rival in the case of water-power being utilised to turn a turbine, which is coupled to the dynamo. In some instances the current is conveyed to a motor many miles away.

In order to avoid confusion in the reader's mind I have hitherto avoided mentioning any but continuous-current dynamos, i.e. those generating a current flowing continuously and uninterruptedly in one direction. There are, however, **alternating-current dynamos**, i.e. those generating a current flowing first one way and then the opposite way. The number of such alternations per second might be, for instance, fifty, so that one can imagine the pulsations of the current could, at such a "frequency" or "periodicity", as it is termed, be hardly noticed. By reducing the periodicity sufficiently,—say by having only ten alternations per second,—the pulsations would be, in many instances, distinctly noticeable to the naked eye. This will be readily understood by imagining an incandescent lamp lighted by a "continuous" current, which would give an uninterrupted light, and then imagining a similar lamp lighted by an alternating current. This alternating current would, as I have explained, first pass through the lamp in one direction and then in the other, and consequently, if the alternations were sufficiently few to the second, there would be a distinct pause at each reversal, and the lamp would at each such moment go out, or have a tendency to do so. For lighting purposes, therefore, a minimum periodicity must always be maintained

An alternating-current dynamo is similar in principle to a continuous-current dynamo, with the exception that the latter has a commutator which commutes the direction of the current, which would otherwise alternate. Turning back to the armature previously described and illustrated in fig. 614, page 239, we note that the wires on the periphery have current induced in them; and as these wires pass the north pole of the electro-magnet, the current is induced in one direction, and as they pass the south pole, in the other direction; thus, were it not for the commutator, the result would be the production of an alternating current.

The commutator, however, has the effect of connecting the wires to the brushes according to their direction, that is to say (turning back to our diagram of the shunt dynamo, fig. 616, page 240, and calling the magnet pole on the right south, that on the left north, and imagining the armature to be revolving in the direction of the hands of a watch when looking at the face), the current in a wire coming, during a revolution, under the north pole, is induced in a direction away from the reader, and is, on reaching the brush, transmitted to the circuit in that direction; then the wire, passing on between the poles of the magnet, becomes electrically idle, and on passing under the south pole has a current induced in it in a direction facing the reader, and this current is transmitted by the second brush (which it has by this time reached) to the circuit in that direction, i.e. opposite to the former direction; this completes the production of a continuous current, since the current from the south-pole brush should flow outwards through the circuit, and thence back into the armature, entering at the north-pole brush, which, as just shown, the commutator has arranged shall have the current going inwards.

Continuous and alternating current dynamos have each their proper sphere of action. The former are by far the more convenient, as will be seen later, but the latter have the distinct advantages of being adaptable for very high pressures, which would not be feasible with the continuous-current dynamo, owing to the difficulties of insulating and commuting currents of more than moderate pressures; they can also be used with transformers, which are apparatus to transform the nature of a current, as I will proceed to describe, without the necessity of having any moving part in such transformer.

I have previously mentioned that electrical energy can be generated at various pressures without altering the value as energy; for instance, one thousand watts can be represented by 500 volts and 2 ampères, by 2 volts and 500 ampères, 1000 volts and 1 ampère, and so on, all of them representing the same amount of electrical energy, namely, 1000 watts.

A **transformer** is merely an apparatus for changing energy in one form, to a similar quantity of energy in another form. Every change so effected must result in some loss, but the loss in changing electrical values in transformers is very small, and so we speak of transformers being of very high efficiency. If we had a dynamo giving 5000 volts and 10 ampères, it would be an easy matter to build a transformer to receive the energy, and give it out in any form we might desire, say, 500 ampères and 100 volts. Such a facility is very useful in electric lighting, as it permits of current being transmitted to a distant point at a high tension, and there changed, by passing it through a transformer, to a low tension.

To explain the advantage of transmitting energy at a high tension and then transforming it, I must first describe the method of conveying it, which I need hardly say is by providing for it a metal path.

The various metals offer different resistances to electricity, and consequently what is required is a metal which offers only a reasonable amount of resistance combined with cheapness. Copper meets the requirements of the case generally better than any other metal, and is, for electric-lighting purposes, universally used. For the comparatively infinitesimal currents used in telegraphy, iron is frequently employed, although its resistance is many times as great as that of copper. Silver offers less resistance than copper, but of course the price is prohibitive. The greater the resistance which the metal offers to the current, the greater is the heat generated by the electrical energy endeavouring to overcome it, and this feature is often made use of, as I have already explained in the section on "Warming and Cooking by Electricity".

As a standard of current-carrying capacity, a square-inch section of copper is recognized as capable of carrying a thousand ampères. This standard is used in all calculations to convey to one's mind the quantity of current, per section of metal, employed in any particular case. An electrician would know that the wires, in a room fitted with the electric light, would be sufficiently large if arranged on the basis of 1000 ampères to the square inch, but he would be at once dissatisfied if he were told that the wires were loaded with current to the extent of 4000 ampères to the square inch, because he would know that, upon such a basis, there would be sufficient heat generated in the wires to damage the insulation covering them, the heat arising from the energy absorbed in forcing the current through such a size of wire.

To return now to transformers, the advantages of which I have to explain. If we transmit 1000 ampères at a pressure of 100 volts for 2 miles, the size of cable must be very large and costly, but if we transmit this electrical energy of 100,000 watts at a pressure of 10,000 volts, we shall only require a cable to carry 10 ampères, instead of 1000, and the cost will be enormously reduced. If it were not for such facilities, the transmission of energy to any considerable distance would be impracticable.

The alternating-current transformer itself is nothing more than an "induction" or "shocking" coil, an apparatus which is frequently seen at fairs and such places, where the proprietor will allow anyone to receive a shock for the sum of one penny. The principle is very simple: If some turns of insulated copper wire are wound upon a piece of iron, as shown in fig. 619, and through this wire (called the *primary* coil) a current is passed, the iron will become magnetic, as previously described. Now if, on top of this copper wire, some more insulated copper wire is wound, called the *secondary* coil, a current can be induced in it from the coil below. In other words, there are two distinct coils wound on a piece of iron, and if we put electrical energy in one, we can get the same amount (less the small losses due to inefficiency) out of the other.

Fig. 619.—View of Transformer.

In each coil, of course, the gauge of wire is arranged to suit the current, and if we require to transform down from a high tension, as is generally necessary in electric lighting, the primary coil would consist of many turns of fine wire, as the current would be small and the pressure high, while the secondary coil would be of thick wire, as the current would be great and the pressure low. For example, if 2 ampères at 20 volts are passed through the first coil, 20 ampères at 2 volts can be obtained from the second, so we shall have obtained a similar amount of energy to that expended, but at a different pressure.

The current on commencing to flow in the primary coil induces a current in the opposite direction in the secondary coil, and, on ceasing to flow, reverses the current in the secondary coil, so that the secondary coil gives an alternating current. If, however, the current in the primary coil were to be continuous and *uninterrupted*, the second coil would not *continue* to produce an induced current; in the shocking coils referred to, the primary coil is fed from a battery which gives a continuous current, but in circuit with it is a little apparatus termed **an interrupter**, and at each interruption, as described above, a current is induced in the secondary coil in each direction, *i.e.* alternating. If the primary coil were fed with an alternating current, the interrupter could be dispensed with, and at each alternation of the primary current, a current would be induced in the secondary coil, also of course alternating in direction.

Such alternating-current transformers give out a buzzing sound, in consequence of the molecular disturbance caused by the current reversals.

Continuous-current transformers, which have the disadvantage of a moving part, and consequently require some little attention, are based upon the principle of the electro-motor.

If a dynamo be connected to a second dynamo by a couple of wires, and the first be driven by an engine, the current it produces will enter the second dynamo and cause it to rotate, when it is called a motor. The reason for this rotating is simple, as the current from the first dynamo or generator enters the armature of the motor, and magnetizes it, but since the magnets of the motor are also magnetized by the same current, the armature is repelled round from the poles of the magnets, the like poles repelling one another. We will suppose that this motor spindle is connected to a third dynamo by a belt. Then this third can be made to generate at whatever pressure is required. So the first dynamo, as a generator, supplies the second, which absorbs its energy and rotates, driving the third dynamo as a generator. If the first were wound for 1000 volts and 10 ampères, the second could be wound to receive such a current, and the third could be wound to generate (say) 100 volts and 100 ampères.

The continuous-current transformer is a combination of the second and third machines, such transformer having one armature, but being wound with two circuits, each with its commutator. One of these circuits absorbs the energy from the first generator, and the other generates at the final pressure required.

We have so far considered dynamos as the chief means of obtaining current for lighting purposes but have not gone into the question of driving them, which is a matter of great importance. There are many **motive powers**, such as the steam-engine, gas-engine, oil-engine, and water-turbine. Of these a selection should be made to suit each individual case. Nothing is cheaper than water-power, provided the first cost of utilizing it is not excessive; after this steam-power is undoubtedly the cheapest, if the steam is required in sufficient quantity and for a sufficient length of time. Gas and oil engines come last upon the list, when the favourable circumstances alluded to above cannot be ensured, but otherwise they are, in many cases, the best means of obtaining power; for instance, where the power is required in only small quantities, and then only occasionally.

Accumulators are very important adjuncts of electric-lighting apparatus. They are not necessarily desirable where electricity can be continuously generated, but it frequently happens that, even if it can be continuously generated, it is at such inconvenience, or causes so much wear and tear, as to justify the use of accumulators.

Fig. 620.—View of Accumulator-cell.

The storage or accumulation of electricity is effected by means of chemical action; and the apparatus employed, in its crudest form, would consist of two lead plates, submerged in dilute sulphuric acid, one connected to the positive pole, and the other to the negative pole, of a dynamo. On a current of electricity being passed from one plate to the other through the liquid, a chemical action is set up, which, on the charging current being stopped, will cause the cell, as it is termed, to give back the energy when called upon. A view of an accumulator-cell in common use is given in fig. 620. A cell, when fully charged, will for a very short time give 2 volts, and when exhausted with working, 1·9 volts. To charge a cell, 2·5 volts are required. The inefficiency of the apparatus is at once apparent, and were it not that this is often over-balanced by the advantage of being able to obtain a small amount of current throughout a long period, just as economically as a large amount for a short period, accumulators would be avoided. As each cell gives only 1·9 volts when discharged, a large number in series are required to form a battery for an electric-lighting circuit.

On charging a cell, the positive plate—that is to say, the plate connected to the positive terminal of the dynamo—becomes the colour of wet chocolate, and the negative plate that of slate. The deeper the chocolate colour, the more thoroughly charged is the cell, and after a little experience, it is an easy matter to decide roughly the amount of charge a cell possesses. The specific gravity of the dilute acid also varies with the charge, and by means of a hydrometer this can be ascertained. Thus there are three ways of determining the charge—*firstly*, by the colour of the plates; *secondly*, by the reading of the hydrometer; and *thirdly*, by the E.M.F. of the cell, which can be ascertained by means of a voltmeter, as will be hereafter explained.

As I have said, each cell gives at the end of its discharge 1·9 volts, so that if an E.M.F. of 100 should be required, 53 **accumulator-cells in series** would be necessary. By the expression "in series" is meant that the cells should be

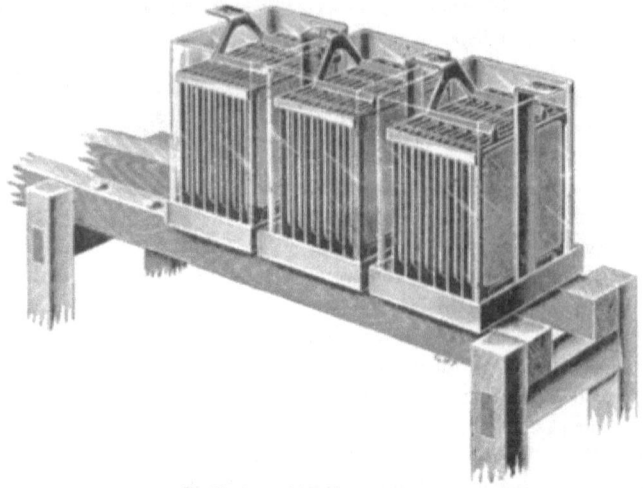

Fig. 621.—Accumulator-cells connected in Series.

joined up in a row so that the current from the end cell must **pass** through the rest—that is to say, the positive of one cell must be connected to the negative of the next, and its positive to the negative of the next, and so on, as shown in fig. 621. But supposing one cell gave 1·9 volts and 10 ampères, the 53 cells connected in series, although giving a pressure of 100·7 volts, would still yield a current of only 10 ampères. If the same number of cells are **connected in "parallel" or "multiple"**, we should have a different result, viz. 53 cells, each giving 10 ampères = 530 ampères, at 1·9 volts. Connecting "in parallel" is merely joining all their positive terminals together on to one wire, and all their negatives on to another. Each cell can then give its ten ampères to the circuit, but the pressure between the two wires is only 1·9 volts, since the voltage has not been, as in the previous case, augmented by the cells being in "series".

As an analogy to explain the above more clearly, imagine three pipes full of water, and of equal dimensions, say 100 feet high and 6 inches in diameter. Now, if we join these "in series", we have a pipe 300 feet high by 6 inches diameter, and we get so many gallons of water with the head of 300 feet, but if we join these side by side, or "in parallel", we only get a third the pressure, but we have three diameters of 6 inches, each giving us water.

Although, in order to avoid confusion, I described cells just now as having each only two plates, they generally consist of many such, as shown in figs. 620 and 621. **The number of plates in cells**, however, is determined merely by the output required. If, for instance, 2 plates give 10 ampères, 4 plates would give 20 ampères, and so on. The number of plates is increased in preference to increasing their size, as the latter method would render them weaker. No matter how many plates there are in a cell, there are only two terminals, all the positive plates being joined together " in multiple" and all the negative " in multiple", usually by means of lead lugs cast on the plates and fused together. These plates are interleaved, so that, on looking into a cell, one sees first a +, then a − plate, then another positive, and so on. There is always, however, one more negative plate than positive, as it has been found that by having a − plate on each side of a + plate, the positive plates, which are the more delicate, are better preserved.

The plates of accumulators, to be commercially useful, must be strong, and must offer large surfaces to the acid, or electrolyte (as it is sometimes termed). In order to achieve this, some of the first makers take a solid lead plate about half an inch thick, and cut or cast fine grooves in it on each side to the depth of (say) one-eighth of an inch; others take a ribbon of lead about half an inch wide by one-thirty-second of an inch thick, and form a plate of about a foot square with it, by bending it back continually on itself.

The surface of the lead plates after charging becomes spongy, and this spongy surface constitutes their capacity; the plate offering the most surface has the greatest capacity. This peculiar surface is generally obtained in the first place by chemical and electrical treatment, and if the plates are used judiciously, the depth of this spongy lead should increase, until, after very many years, the entire plate should become spongy and the cell attain its maximum capacity, after which it will fall to pieces. The procuring of such a spongy surface by pasting prepared lead on to the plate, is obviously a less substantial method than the production of the spongy surface from the plate itself.

The selection of a cell can be easily made with confidence, if the two points I have mentioned are kept in view, viz. a maximum of strength and a maximum of surface. Few, if any, makers exceed the size of about 12 inches square for their plates.

In stationary work in small installations, the plates are set up in glass boxes, which permit of the plates being readily examined; but in electricity-supply works, where the cells are very large, or in cases where they are subjected to rough treatment, such as in tramway or yacht work, lead boxes are used with an outer casing of teak.

The connections from cell to cell, in small installations, are generally made with brass bolts and nuts, clamping the lugs together tightly. All are then varnished to protect them from the acid. In large installations, such as electricity-supply works, these lugs are often burned together with lead.

The capacity of cells is described in ampère-hours, that is to say, so many ampères for so many hours; for instance, if a set of cells is described as having 600 ampère-hours at 100-volts pressure, it would be understood that, at that pressure, 600 ampères could be obtained for one hour, or 300 for 2 hours. I mention this figure to make the point clear, but have now to add that the 600 ampère-hours (or A.H., as it is often written) are conditional on a maximum discharge of (say) 60 ampères not being exceeded, so that the quickest method of discharging the cells permissible, would be 60 ampères for 10 hours. The voltage of 100 would only indicate the *number* of cells, since one cell giving a minimum of 1·9 volts would discharge at 60 ampères for 10 hours, and fifty-three in series would also discharge 60 ampères for 10 hours, but (as stated above) at a pressure of 100 volts.

The voltage of 1·9, so often mentioned here, is only that to which makers recommend their cells being discharged. In some instances, a figure as low as 1·8 is stated to be that to which a particular make of cell will discharge, but as it is only a question of how low the cell can be discharged without injury to it, one can understand it is not advisable to "cut it too fine".

The charging of cells, unless efficiency is of great importance compared with durability (as is seldom the case), is continued at a maximum current decided by the makers, until the cell "boils",—in other words, until the electrolyte or acid gives off gas so freely as to give the cell that appearance. The temperature of the cell is not to be thought to have increased by the so-called "boiling" of the electrolyte. On the appearance of "boiling" in all the cells, the charging is discontinued, and the total E.M.F. will then, if the cells are healthy, be generally found to be represented by 2·2 volts per cell. With glass boxes for the plates, the "boiling" will be found to give a "milky" appearance to the acid, until the charging ceases.

Much spray is given off during the "boiling", which occurs fortunately only towards the end of the charging, and is a sign that the plates are approaching the desired condition. To prevent this spray, many devices have been thought of. One is to lay a piece of glass over the top of the glass box, in order to catch and throw back the spray. To effect this, the glass should be slightly raised at one end, so that the spray accumulating thereon may drop back into the box. Care must be taken that these glass plates do not touch both the negative and positive plate in any box, as, should they do so, a path would be formed for the current from the negative to the positive, apart from its legitimate circuit. To show the objection to this, it is only necessary to remember that a cell consists of a number of positive and negative plates interleaved with one another, but separated from each other with scrupulous care, the only connection being by means of the liquid. If something were to connect the positive and negative terminals outside the box (as the wet sheet of glass would do), the cell would have a circuit from (say) the positive plate through the liquid to the negative plate, and thence along the wet sheet of glass back to the positive plate.

Another method is to fill the glass box within about half an inch of the top with the liquid,—usually the liquid need only be so high as to well cover the plates,—and then to pour upon it hot paraffin wax of the best quality to a depth of the remaining half-inch, having previously fastened in one corner, for the moment, a greased wood plug, which can then be removed, leaving a hole in the wax of half an inch in diameter, through which a little acid can be withdrawn by a syringe. This syringe, or any such tool used with accumulators, must not be of metal, but of glass or other insulating material, for fear of accidentally connecting the positive and negative plates, which would have the effect of the wet sheet of glass above referred to, but with far more disastrous result.

This arrangement of wax is very effective as regards preventing the spraying of the acid, and if the level of the liquid is reduced with the syringe sufficiently to allow the gas formed to escape through the hole made by the temporary plug, all will be well, but the plan has the decided disadvantage of preventing the removal of the plates, or the giving of attention to them to dislodge any scale forming between them (as is often found necessary), without the removal of the wax.

Another plan, and one which I consider by far the best, is to allow to float on the top of the acid a quantity of prepared crumbled cork, or small glass balls. These are easily removed, and neither of them can do harm, provided the cork is prepared to withstand the action of the acid.

Anything which will carbonize and become water-logged is, of course, objectionable, as in time the effect might be, that enough would sink to build up a connection between the plates, and so bridge across the positive to the negative.

To read the E.M.F. of the cell, as mentioned on page 252, **hand-voltmeters** are often employed, which are generally fitted with a piece of double flexible wire and a "spear", as shown in fig. 622. The spear is about a foot long, and the free ends of the attached wires are provided with blades, one of which engages in the positive and one in the negative plate, so as to ensure a good contact whilst the reading is being taken on the voltmeter. The contacts with such a spear are, however, very treacherous, owing to the oxidized condition of the surface of the lead strips leading to the plates in the cell,

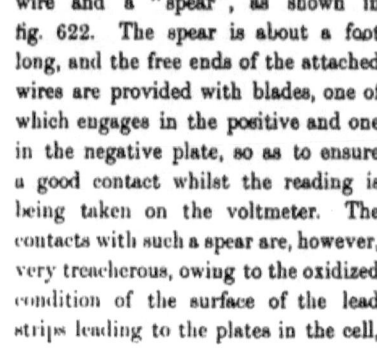

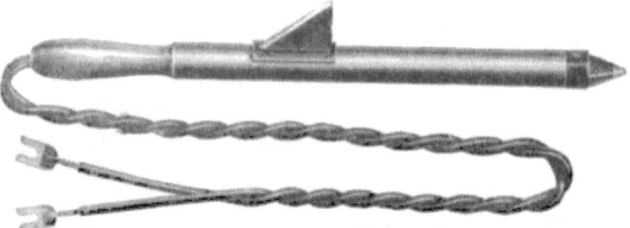

Fig. 622. Hand-voltmeter and spear.

and care should be taken to make the blades thoroughly scratch the lead, or an incorrect reading will be the result.

These hand-voltmeters are only suitable for making comparisons; for instance, if all the cells but one read about 2 volts, and this one only reads 1·5 volts, there is no doubt of the cell being faulty. The hand-voltmeter is quite likely to read to 3 volts per cell, but one must not be led to believe the cell is giving that E.M.F., but merely to use the figure for comparison with the other cells. Above all, the E.M.F. of a cell should not be taken, and the figure multiplied by the number of cells; for instance, if the cell above-mentioned should read 3 volts on the hand-voltmeter, and there should be 50 cells, it should not be concluded that the voltage is 150. On the contrary, no reliance should be placed upon such a figure.

The principle of the voltmeter is very simple, and one form, which will suffice to explain it, is that of an electro-magnet attracting a small piece of iron. The greater the strength of the current, the greater will be the quantity of current flowing round the electro-magnet, and the greater will be the attraction of the magnet for the small piece of iron; consequently, the greater the attraction. the greater distance will the indicating needle move, to which it is fixed.

A voltmeter, being designed to measure pressure, must be connected across the terminals of the article to be measured; for instance, to measure a dynamo, one of the terminals of the meter must be connected to one of the dynamo-terminals, and the other terminal to the other dynamo-terminal. In the same way, to measure the E.M.F. of a cell, one terminal must be connected to one plate, and the other terminal to the other plate.

The only reliable method of finding the total voltage (that is, the voltage of all the cells in series) is to connect them to the engine-room voltmeter, which is a similar apparatus to the hand-voltmeter, except that the latter is a small portable instrument with spear, reading to about 3 volts, whereas the former is a much larger instrument, screwed to the switchboard (presently to be described), and with a scale reading to (say) 140 volts.

An ampère-meter (or ammeter) is an instrument on the same principle, except that whereas a voltmeter is bridged across the terminals of the apparatus to be measured, and has consequently to be wound with many turns of fine wire to permit only just enough current to pass through it to move the indicator, an ammeter has to be placed in the circuit, the current which it is desired to measure passing through it. If it were required to ascertain the current given out by a dynamo, it would be necessary to connect one terminal of the dynamo to the ammeter, and the other terminal of the ammeter to the circuit, or, figuratively speaking, to cut the circuit and insert the ammeter.

Both voltmeters and ammeters must be used for such circuits only as they are intended for. If the 3-volt hand-voltmeter were connected across a 200-volt dynamo, it would be burned up with the excessive current passed through it, as it was only wound with sufficient wire to withstand a pressure of 3 volts. In the same way, if an ammeter with a scale of from 5 to 20 ampères were put in a circuit carrying 1000 ampères, it too would be burned up; but if it were put in a circuit carrying ·01 of an ampère, it would not indicate anything, since its scale commences at 5 ampères. The very low readings of long-range instruments are always treacherous; so on a 150-scale voltmeter, no reliance should be put on the readings of (say) the first ten volts.

CHAPTER IV.

ELECTRICITY: WIRING AND LAMPS.

"**Wiring**" is a comprehensive term, including all methods of conveying electricity through wires to and from the various points at which it is to be utilized. Before the pros and cons of the different methods can be discussed, it is necessary to understand the fundamental principle, that the electric current requires a circuit or path, from the point at which it is generated, to the point at which the work is to be done, and thence back to the original point at which it was generated.

The size of the wires must be in proportion to the current conveyed and the distance traversed, in order that very little of the energy may be consumed in them, leaving nearly all the original energy to be usefully absorbed in the work in view; for instance, if we take an incandescent lamp,—which is merely a thin strip of carbon in a vacuum, preserved by a light glass globe,—it is in that strip of carbon (or filament, as it is termed) that the energy should be absorbed; so the wires leading to it should offer very little resistance to the current, and the pressure be such as will force the required amount of electricity through the high resistance of the carbon to make it incandescent, and so give light.

The pressure of the current should be decided upon according to the area or distance through which the current has to pass. In domestic lighting by electric incandescent lamps, such pressures as 60, 100, or 110 volts are usually employed, because these have been found convenient for such work. That the theory, however, may be correctly understood, I would point out that 5, or 10, or any other number of volts, would be equally effective, as will be seen by the following explanation. It has been already explained that a given quantity of electrical energy can be divided into whatever pressure and current may be found necessary,—that is to say, 1000 watts of energy is represented alike by 100 volts and 10 ampères, 500 volts and 2 ampères, or any other combination the product of which is 1000. If it is decided to use an incandescent lamp having such a resistance of filament that 60-volts pressure is necessary to force one ampère through it, that lamp would be said to require 60 watts, and if the one ampère is found to make the filament hot to such an extent as to emit a light of sixteen candles, the sixteen-candle-power lamp is said to take 60 watts. The filament of the lamp, however, might be of such reduced size that half an ampère, on being forced through it by the 60-volts pressure, would be sufficient to make it hot enough to give the same light of sixteen candle-powers.

To understand this proportioning of the filament, it is only necessary to imagine a lamp with such a large diameter of filament that the one ampère would not render it hot at all, but would pass easily through it, just as it does through the copper circuit-wire, in which, of course, it does not appreciably raise the temperature. The length, however, of a filament of such an increased diameter would have to be extended, if it were required to offer sufficient resistance to the 60 volts to permit only one ampère to pass.

From these two examples, since an equal amount of light is obtained in each case, one would be apt to say that it would be more economical to take the smaller amount of current, and this would undoubtedly be true as far as first cost is concerned, but the smaller filament has to be raised to a higher state of incandescence to make it give the same light, and it is consequently not likely to stand the wear and tear so long. By actual experience, a standard has been fixed, at which it is reasonable to work filaments.

A **sixteen-candle-power lamp** is considered to have a sufficiently long life when absorbing 60 watts, at which amount of energy the filament is not unduly heated, and it will be seen that a light of this intensity, absorbing a total of 60 watts, absorbs for each candle-power a little under 4 watts, which is the method used to describe the light-giving efficiency required;—that is to say, a lamp-manufacturer, offering lamps of sixteen candle-powers at 4 watts per candle, states directly the class of article, leaving the purchaser to order them at 100 volts or 60 volts, according to the pressure of his source of current.

At the present day a **light-giving efficiency** of one candle-power for every $2\frac{1}{2}$ watts consumed, which would be equivalent to a sixteen-candle lamp absorbing 40 watts, is considered very high, but, by improvements in manufacture, the light-giving efficiency of lamps will no doubt be from time to time increased, the great problem of production being to find a process to render a filament so strong as to stand an increased state of incandescence, by which it will give more light for the same energy expended. The expression "light-giving" efficiency has been used to prevent confusion with "life" efficiency, which is the length of time the filament will bear being raised to the stated degree of incandescence.

It is an easy matter for a lamp-maker to guarantee any light-giving efficiency, but when this is coupled with life-efficiency, it is another matter; for instance, the 60-watt lamp could be made to give a light of (say) 200 candles, but only perhaps for a few minutes, as the filament could not stand such a very high degree of incandescence much longer. A 60-watt lamp is generally supposed to stand burning for 1000 hours, and a 40-watt lamp perhaps only 600 hours.

The efficiency in watts and life, then, are important points to stipulate for. All makers of incandescent lamps, I believe, refuse to guarantee any particular life, as the article is a very treacherous one to make, and they are unable to feel satisfied that the lamp has received proper treatment.

To show how a lamp may be improperly treated, and at the same time to help to explain the above matter, we will suppose we have a 100-volt circuit, and put a 60-volt lamp in contact. The result will be an intense light, since the lamp has been made to take (say) one ampère with a pressure of 60 volts, whereas the increased voltage of 100 will send through it far more than one ampère, and will raise the degree of incandescence much higher than that for which the filament was intended. This would probably have the effect, during the short time the lamp could stand such treatment, of permanently blackening the inside of the glass by the volatilization of the carbon.

Except by a mistake, no lamp is likely to be so badly treated as the one mentioned, but it is of daily occurrence for lamps made for (say) a 100-volt circuit, to be submitted to 105 or 110 volts for a short time, owing perhaps to imperfect governing of the engine driving the dynamo, thereby allowing the latter to run too fast and so give too great a pressure.

The connection of incandescent lamps to the circuit will now be described. We will suppose that we have a compound-wound dynamo of 100-volts pressure, which, of course, being compound-wound, will, at a constant speed, give a practically constant pressure, irrespective of the load, and we will further suppose that we have some 16-C.P. and some 8-C.P. incandescent lamps. Immediately we connect the two ends of the filament of a 16-C.P. lamp, which protrude through the glass, to the two terminals of the dynamo, we have provided for it a complete circuit, and the current will flow through the lamp and render the filament incandescent. There need be no doubt as to the result, since we have a 100-volt circuit, and have attached a lamp constructed to allow such a current to go through it, at a pressure of 100 volts, as will give a light of sixteen candles, the current in this case being ·6 of an ampère, so that the lamp takes (1000 × 6 =) 60 watts.

We will now hang another lamp on, in exactly the same manner. It also will immediately light, and we shall have two lamps, each giving a light of sixteen candles, and taking ·6 ampère at the common pressure of 100 volts, since each offers a distinct and independent path for the current from the dynamo. For every lamp we so attach, the dynamo will produce just sufficient current and no more.

If we now in an exactly similar manner, and as if the other lamps did not exist, attach an 8-C.P. lamp, we shall have two of 16 and one of 8 C.P. burning simultaneously, the last-mentioned taking, if it is a 30-watt lamp, only ·3 of an ampère, in consequence of being of less power.

We can continue to attach lamps until so much current is required from the dynamo that the wires on its armature become excessively hot, and the cotton, or other insulation employed on it, is impaired. I put it in this way to make it clear, but of course a dynamo is known to have a certain capacity, such as 100 volts and 50 ampères, which would not in practice be exceeded.

Just as the lamps were attached one by one, they can be disconnected one by one; if we wish to extinguish any particular lamp, it can be disconnected without affecting the others.

These lamps, it will be noticed, are connected to the dynamo in "multiple" or "parallel", an expression which has been already explained.

In practice it would be very inconvenient to have a bunch of lamps attached to the terminals of a dynamo, but if **the copper wires for each of the lamps** were many yards long, the dynamo could be in one position, and the lamps as far off, from it and each other, as desired. The copper wires, as already mentioned, should be of such a size as to absorb practically none of the energy. Two main wires or cables, insulated suitably, are attached to the dynamo-terminals, which are usually two copper screws, and the wires are thence carried through the building as required. To any point along the wires lamps can be attached with the same effect as if they were attached actually to the dynamo-terminals, provided that these mains are of such a size as not to absorb any appreciable quantity of energy.

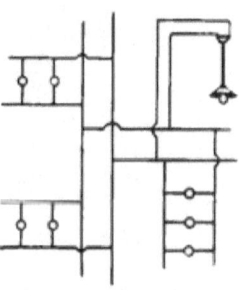

Fig. 622.—Diagram of Tree Wiring.

This arrangement can be elaborated by connecting at any distance from the dynamo along the two cables, two more cables, one to each of the first, and then carrying these two (to which lamps can be attached) in another direction. In the same way, from any point on the two latter, a further pair of cables can be taken, and so on. This arrangement is called **the "Tree" system of wiring**, and is illustrated in fig. 623; the current from either terminal of the dynamo through any lamp, and back to the other terminal, must, of course, be uninterrupted.

As it would be most inconvenient to detach a lamp, when the light from it is not required, an arrangement called a **switch** has been designed to give the

Fig. 624.—View of Small Switch.   Fig. 625.—View of Small Switch with the Outer Case removed.

same effect. It does nothing more than break the circuit to which it is attached, and so prevent the current passing. Cutting the wire, either before or after it passes the lamp, would serve the same purpose as the switch, but the switch has the advantage of severing the connection by a simple piece of mechanism, the

Fig. 626.—View of Double-pole Main Switch.

obvious part of which is simply the rotation of a button, or pressure upon it. A switch can be so located that it will cut a wire feeding one lamp, a dozen, or any other number, and to make this clear, I may mention that at least one of the wires leading from the dynamo is always passed through a switch within a few feet of the machine, that by its use all the lamps can be extinguished or lighted. I say "at least one wire", because the best method with a large current would be to have a switch on each main cable, close to the dynamo. These two switches are in such a case combined on one base, and controlled by one handle, when they are called a double-pole switch, since they break both electrical poles of the circuit.

As it would be inconvenient to attach lamps by the small wires protruding from their glasses to copper wires in order to connect them to the main cables, "**lampholders**" are provided, which vary considerably in design, but which do nothing more than hold the lamps on to the wires leading to them. One variety is illustrated in figs. 627 and 628, and a lampholder combined with a switch in fig. 629.

We have now in our mind a number of lamps connected at various points throughout the length of the main cables and their branches, and we know that each lamp is allowing only a little current to go through it, in consequence of its high resistance; but suppose that, by some mishap, the two small copper wires supplying a lamp were to become connected together, then this path would offer a resistance to the dynamo very much less than all the lamps put together, because, as so frequently mentioned, the copper mains and wires have been arranged of such a low resistance as to absorb practically no energy.

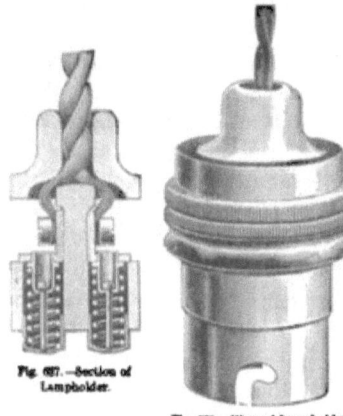

Fig. 627.—Section of Lampholder.

Fig. 628.—View of Lampholder with Outer Case.

Fig. 629.—View of Switch Lampholder.

This fault is termed a "short circuit", as the current has found a shorter or easier circuit than its proper course.[1] The result would be that all the current would go through these two wires at the point at which they came in contact, and none through the lamp, which would therefore be extinguished. Further, the two small wires, which come in contact and make the short circuit, will receive all the current the dynamo is producing, whereas they were only arranged to supply one lamp taking perhaps ·6 of an ampère, and these small wires, becoming exceedingly hot, will burn the rubber, &c., with which they are covered, and this may lead to a serious fire in the building.

This danger, however, can certainly and easily be avoided. If it were not so, electric lighting would not be, as is universally admitted, the safest known artificial illuminant. The cure is based on the fact that lead and tin wires have low melting points, and sufficiently low electrical resistances to enable the following arrangement to be adopted. At the point where one of the two

---

[1] The expression "easier circuit" is less misleading,—although never used,—as it is not a question of distance but of resistance.

small wires, leading to or from a lamp, leaves the main cable, or sub-main cable (as a large branch is termed), this small wire is cut, and the two ends so produced are connected by (say) an inch of lead or tin wire of perhaps 20 B.W. gauge, so that the lamp will light just as it did previously, but the current has to pass through this inch of tin wire on its way to or from the lamp. This small piece of tin wire does not offer appreciably any increased resistance to the current, but unlike the copper wire to which it is attached, it cannot carry much more current than the ·6 of an ampère without melting.

Where this method is adopted, and a short circuit happens, the current increases instantly along this small circuit to a very large quantity, and in less than a second the tin or lead wire melts, and automatically switches off the faulty circuit. Nothing could be more simple or effective. By adding one of these pieces of tin wire at each branch, and also in the main cables close to the dynamo, we make it absolutely impossible for any circuit to take such an excess of current as will raise the temperature of the copper cables and wires more than a few degrees, since any faulty circuit will be immediately and automatically cut out.

The article used for preventing this overheating of the wires is termed a **fusible "cut-out"**, and consists of any shape of box (preferably of china) with small terminals within it, by which to attach the copper wires to the small piece of tin wire. The gauge of the tin wire would have, of course, to be varied to suit its position—that is to say, if the cut-out were fixed in the main cable near the dynamo, the tin wire within it would require to be much larger than the small-gauge piece for conveying current to one lamp.

Fig. 430.—A Fusible Cut-out.

Incandescent lamps are often suspended by two flexible wires, twisted together in the form of a cord, which enters a small fitting on the ceiling. This fitting is called **a ceiling-rose**, and is usually made of china. Its object is to provide terminals for the two small copper wires from the main cables, which have to be brought to that point, and also terminals for the double flexible wire referred to. In this fitting there is also a fusible wire, so that, on any fault happening (in the flexible cord, for instance), the entire length of this cord would be automatically cut out of circuit.

Mention was made just now of cut-outs being inserted near the dynamo, and to explain the importance of **the position of cut-outs**, we will suppose that two cut-outs are fixed, one on each cable, leaving six feet of cable between the dynamo-terminals and the cut-outs. If a fault arose in this distance, there would, of course, be no fusible wire to rely upon, but as such cables are always most securely fixed and strongly insulated, they are practically free from the danger in question. Such is not the case, however, with smaller wires; so, if the cut-outs are not fixed very closely indeed to the cable, from which the wire leading to them is connected, the danger exists of a short circuit happening *behind* them.

If a short circuit did happen, there would be the main and sub-main cut-outs to rely upon, but nevertheless electrical engineers consider the point of such importance that they often fix the cut-out actually on the top of the cable or wire from which the

Fig. 631.—A Ceiling-rose.

Fig. 632.—Wall-socket and Plug.

branches are being taken, so that it is literally impossible to leave any length of wire "behind" it unprotected.

There is another appliance frequently used in the lighting of rooms, namely, a "**wall-socket**" or "**connection-socket**". This consists of a wall-piece and a plug, as shown in fig. 632. The wall-piece contains two small terminals inclosed in a small china case, and connected through a fusible tin wire to the mains; the plug also has two terminals, which fit into two holes in the wall-piece, and so connect with the two terminals therein; the two terminals in the plug are connected to a double flexible wire, which may be attached to a lamp or other fitting. In this way, the flexible wire can be put into contact with the mains simply by inserting the plug into the china cover; a portable reading-lamp, hand-lantern (see fig. 633), or any such fitting, having a piece of flexible wire attached, can thus be readily connected to the mains, and so provided with current.

The atmosphere is fortunately an insulator of electricity, although varying in value considerably according to the amount of moisture suspended therein. If we take an incandescent lamp and attach one terminal of it to one main cable of a circuit, supplied at the time with a potential of (say) 100 volts, and then attach to the other terminal of this lamp a short piece of wire, and with the other end of this short piece of wire approach the other main cable within (say) a one-hundredth part of an inch, the intervening space of air (only one-hundredth part of an inch) would be sufficient to prevent the electricity from jumping across and lighting the lamp; it would, indeed, need a very much higher potential to do so than any used in domestic lighting. If the wire were brought closer and closer until it actually touched the main, contact would be made without a spark occurring. But if the wire were then removed, the potential already existing in it (which did not exist before the contact was actually made) would be sufficient to bridge across a short space of air.

Fig. 133. Portable Hand-lantern.

It will be seen, then, that it is not the making of a contact, but the breaking of it, which produces a spark. In consequence of this, **switches made with a trigger action** are now used, by which the lever, which makes and breaks the contact, is in the latter case very suddenly withdrawn without the control of the operator. In other words, if one were to turn a lamp out with such a switch, it would be found that, as the contact was about to be broken by one's hand turning the button, the lever inside would suddenly and quickly make the disconnection, even if one tried to make it do so slowly. These switches are termed "quick break". There are also "quick make" switches, but these are not so necessary. The advantage of the quick break is, that the spark is not allowed to exist for any greater length of time than is absolutely necessary; thus the metal parts, which would otherwise suffer somewhat, are practically undamaged.

We have hitherto spoken of **insulated copper wire** in a general way only; before proceeding further, it would be well to understand that the insulation of wires not only consists of cottons, silks, and other materials used in the manufacture of dynamos, &c., but also of a complete coating of vulcanized rubber, giving a result somewhat similar to the placing of a cable inside a garden-hose.

The advantage of this is that the wire can be subjected to any amount of moisture, without this being able to pass through the insulation. This is of great importance, because moisture is a conductor of electricity. If a pair of wires get into a pool of water, or be placed on a damp wall, and be imperfectly insulated, a current will pass from one to the other, and the short circuit already described will in time occur.

The quality of the insulation is selected to meet the requirements of each case, but it may safely be assumed that, for domestic lighting, the insulation known as 600-megohm quality is sufficient. The expression "600 megohm" merely means that, with a high E.M.F. of (say) 400 volts (far in excess of what would be used in buildings), the insulation in question would offer a resistance of 600 million ohms,—a megohm being equal to one million ohms.

The "tree" system of wiring so far described, with its cut-outs at the branches, is only used in mills and such places, where the following system would be too costly and impracticable. It will be noticed that, in the tree system, for each lamp there must be at least two joints and one cut-out, or, if double-pole cut-outs are used, two cut-outs. The joints must, of course, be perfectly insulated.

Insulated joints are made by baring the ends of the copper wire, rolling these ends together, soldering the joints, and afterwards insulating them with a strip or ribbon of rubber, and then with waterproof tape. For a flux with the solder on all joints, resin only should be used. These insulated joints are by no means equal to the 600-megohm wire, as regards their ability to resist moisture; excellent and costly wire is therefore much impaired by having its insulation cut in many places, and replaced with something of a comparatively worthless nature.

To avoid this, **the "distributing" system of wiring** has been largely adopted. It consists in taking the mains from the dynamo to two terminals on a slate slab, and on that slate slab branching these two points into (say) 12 pairs of points; then from these 12 pairs of points, taking sub-mains to twelve more slate slabs in different parts of the building, from each of which a number of circuits could radiate according to the number of lamps to be lighted. No joints are required at all with this system, and, as all the cut-outs are upon the slate slabs only, any particular cut-out is readily found and replenished, whereas with the "tree" system in an intricate building, it is often a game of "hide-and-seek" to find a cut-out when a lamp has gone out. If the slate slabs, or distributing-boards, as they are termed, are only in dry places, the wiring can be subjected to a very considerable amount of moisture before being damaged thereby.

Since an arrangement like that just described would be very costly for a large number of lamps in an intricate building, **a compromise between the two systems** is generally arrived at, by which, from the distributing slate slabs, circuits radiate to a number of lamps, each taking not more than ·6 of an ampère. For these lamps, joints are allowed as in the "tree" system, but cut-outs only upon the distributing slabs. This means that the number of joints is kept down considerably, and the cut-outs readily found, especially if each pair (for in such a system one would be upon each wire) is inscribed with the number or description of the lamps they protect.

All wires, except flexible pendants and those supplying portable lamps, should be protected by being laid in **hard-wood casing**, consisting of a strip of

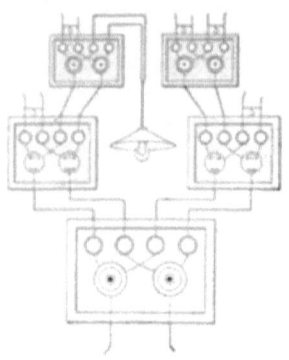

Fig. 634.—Diagram of the Distributing System of Wiring.

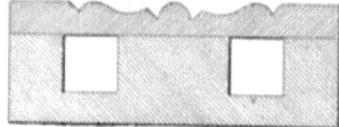

Fig. 635.—Section of Wood Case and Lid for Electric Wires.

wood having chased in it two grooves, according to the size of the wires to occupy them, and having, when the wires are placed in position, a thin strip of wood screwed on the face of it to form a cover. In damp positions, as in cellars, the casing should be varnished inside and out, back and front, with at least one coat of shellac. A substitute for wood casing has recently come into vogue, and consists of tubing formed of paper, rendered waterproof with bitumen or similar substance. Wrought iron is often employed in works where the treatment is of the roughest kind, and the moisture excessiv

We have so far considered only incandescent lighting, but **lighting by "arc" lamps** is another form, and of very great value. In describing the breaking of a circuit with a switch just now, whereby a spark was created, it would have been quite correct to have termed such a spark an arc; any space over which electricity passes is, in electrical parlance, an arc.

With incandescent lighting, the electrical energy is, as has been shown, consumed in passing through a filament of carbon; but similar energy can be consumed in "arcing" between two points. If a piece of carbon is held in each hand, each piece being connected by a wire to one of the mains of the circuit, and these two pieces of carbon are brought into contact, current will begin to flow, and on drawing them slightly apart the electricity will be able to maintain an arc across the space dividing them, and, while doing so, will render the points incandescent and so give light.

There is no other arrangement known, which, for the same amount of energy, will give so much heat as the electric arc, and it is to this that the high state of incandescence is due. It has been seen that 60 watts give, with an incandescent lamp, about 16 candle-powers, but as two arc-lamps, taking (together with the resistance) 100 volts and 10 ampères, or a total of 1000 watts, will give about one thousand candle-powers each, their great comparative efficiency can be at once appreciated.

The carbon points, not being in vacuum (as are incandescent filaments), are slowly consumed, and the pieces of carbon must therefore be gradually fed together, to keep the distance between them constant. Carbon is used in preference to any other material, owing to the high state of incandescence to which it can be raised, and to its lasting properties. The arc has the peculiar action of forming the end of the positive carbon into the shape of a small cup, while the end of the negative carbon is brought to a slight point, which will approximately fit into the positive cup.

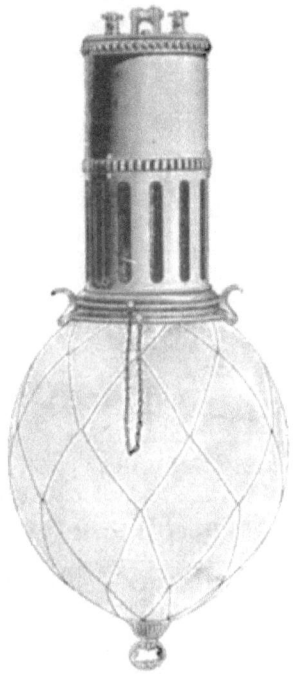

Fig. 696.—An Arc Lamp

The space between the carbon points must be regulated according to the current of electricity, otherwise a white and steady light will not be obtained. With an arc of ten ampères, which is a very usual quantity, the distance apart of the two carbons to obtain this white light should be about one-eighth of an inch; but ten ampères passing through an eighth-of-an-inch arc only require about 45 volts, so if we were to put such an arc on a 100-volt circuit, the potential would be too great, and more than 10 ampères would pass. To prevent this, two such arcs are used in series, that is to say, the current from one main has to pass through one arc and then through the other, before it is allowed to reach the other main. This gives a total of 90 volts absorbed out of the 100 provided by the circuit. The balance of 10 volts is therefore consumed by a resistance placed anywhere within this circuit.

This arc-lamp resistance usually consists of a length of iron wire, wound into the shape of spiral springs so as to occupy as little space as possible, the springs being inclosed in a cast-iron box, as shown in fig. 637. A resistance, like every-

Fig ... Arc-lamp Resistance

thing absorbing electrical energy, becomes hot when in use, and is therefore mounted on a slab of slate, and packed out from the wall to which it may be fixed, so as to allow a current of air to pass behind it, and so prevent the heat affecting the wall. When the voltage of the circuit is (say) 110, the resistance must be increased to absorb a further 10 volts.

It is not advisable to run two such lamps on less than a 100-volt circuit for the sake of avoiding the waste in resistance,— which, it will be noted, represents ten per cent of the total energy,—since the resistance has the beneficial effect of steadying the current passing through the lamps, and consequently the light given by them.

The light from arc-lamps has been described as white, and so much is this the case, that colours can be matched thereby with the same readiness as by daylight, a fact which drapers and others are not slow to avail themselves of.

The mechanism attached to each pair of carbons, it would at first appear, should be on clockwork principle, and arranged to feed the carbons together at a speed which will exactly meet the rate at which the points are consumed. In the early days of lighting such an arrangement was used, but it was soon found that the carbons, varying in quality throughout their length, required some means of feeding them together at varying speeds. This has been achieved electrically by taking advantage of the following principle. If the carbons (which, at the commencement, were one-eighth of an inch apart, and had passing across them 10 ampères requiring a voltage of 45) were not moved, they, on burning away, would shortly require more than 45 volts to maintain the 10 ampères, or, in time, to maintain any arc at all, so, on the difference of potential or voltage between the two points increasing to above 45, an electro-magnet as a by-pass automatically obtains sufficient current to start the mechanism and force the carbons together to the required distance.

The peculiar cup-shape of the positive carbon, called the "crater", throws the light directly in front of it, so that if the two carbons should be placed vertically, and the positive or cup-shaped carbon be uppermost, practically all the light will be thrown downwards. By reversing the current without affecting the position of the carbons, that is to say, by making the lower carbon positive, all the light is thrown upwards. The former, shown in fig. 638, is the arrangement generally seen, with a globe of opalescent glass inclosing it. The latter, shown in fig. 639,

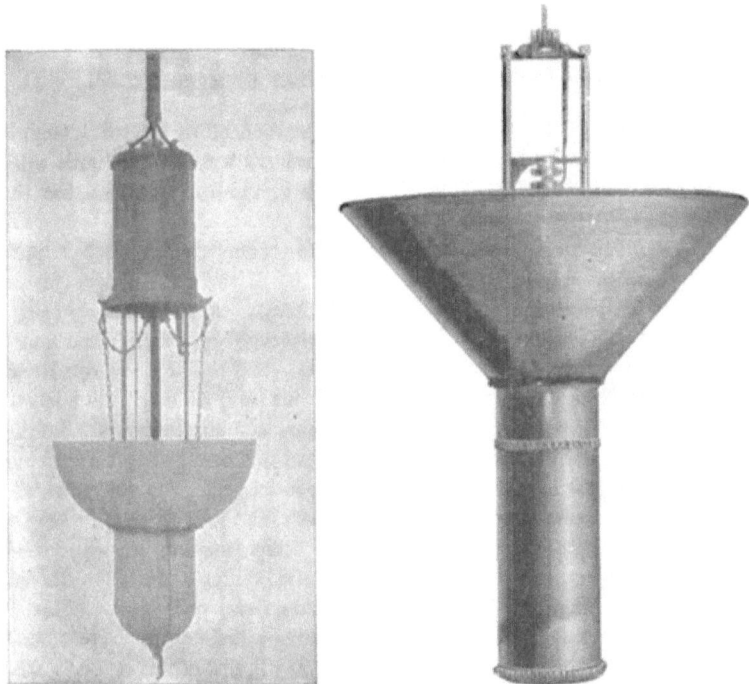

Fig. 638.—Arc-lamp with Half-globe Reflector.   Fig. 639.—Inverted Arc-lamp.

is called an inverted lamp, in which the carbons are merely shielded from the eyes of those below by a metal or thick glass cone, so that the source of light cannot be seen, but the rays therefrom proceed to the ceiling, whence they are reflected to the floor. Given, then, a good white ceiling, the light thrown down is as near an approach to daylight as has yet been found, since there is no brilliant spot, from whence it proceeds, to irritate the eyes.

I stated just now that an **arc is not formed in vacuo**, as are incandescent filaments. But arcs have been used, and are likely to be more and more frequently used in a partial vacuum, whereby the consumption of carbon is decreased and the light rendered softer, although somewhat at the expense of purity of colour.

CHAPTER V.

ELECTRICITY: AN INSTALLATION IN A HOUSE.

We will now consider ourselves in the position of an electrical engineer, designing and supervising **the lighting of a country house.** This work will, of course, offer a great contrast to an installation in a mill or workshop, but these buildings are beyond the scope of this book.

The first point to be considered is **the number of lamps** which will be required. Unless there are unusual conditions, the capacity of the accumulators (see page 251) must be two-thirds the number of the lamps. As each lamp should for this purpose be considered as of 16 C.P.,—although eventually some may be increased and others decreased in candle-power,—and as, on a 100-volt circuit,[1] each lamp requires ·6 ampère, the total current for (say) 66 lamps will be 39·6 or (say) 40 ampères, which, for 10 hours, will equal 400 ampère-hours. Sufficient cells will be required to give a pressure of 100 when discharged to 1·9 volts each, at which point the makers advise their being recharged. The 100 volts, then, divided by 1·9, shows that 53 cells are necessary, and as it is usual to keep one extra in circuit, as a reserve in case of repairs being necessary, we will order 54 cells in glass boxes, each of 400 ampère-hour size, and at a maximum discharge of 40 ampères, complete with glass oil-insulators and teak trays. These oil-insulators are merely small glass cups, containing a little oil to prevent surface leakage (moisture being unable to creep over the oil), upon three or four of which is placed a shallow tray filled with sawdust or crumbled cork, so that the glass box of the cell (the bottom of which is often very irregular) will, when placed on it, be evenly supported.

Where gas is not obtainable, **an oil-engine** must be used, but a gas-engine is preferable, owing to its greater simplicity and reliability. To ascertain the power of the engine required, we must consider the current needed for the accumulators again. This is 40 ampères, and as each cell on completion of charging requires

---

[1] I have fixed the E.M.F. at 100 volts, as I am imagining a case in which there are some outbuildings, about eighty yards from the house, where we can install the plant without any noise from the engine or smell from the cells causing annoyance; such a distance necessitates rather a high E.M.F. to avoid having very large mains.

2½ volts, that will be 54 cells multiplied by 2½, or a total E.M.F. of 135. These volts multiplied by the 40 ampères will give a result of 5400 watts, and, as 600 watts to the horse-power is a safe figure to estimate upon, we divide the 5400 watts by the 600 and find that an engine of 9 horse-powers is required.

We shall then order a 9 brake-horse-power engine, being careful to use the work "brake", since it is 9 *actual* horse-powers imparted to the dynamo that are necessary, and not 9 *nominal* horse-powers, which is obviously a vague expression, or 9 *indicated* horse-powers, since, although the engine may indicate a certain power, this is not definite enough, as a considerable proportion of that power is expended upon the compression of its gases, &c. What we want is a 9 brake-horse-power engine, and of such a speed as will enable it to run without imparting the power to the dynamo in pulsations. This we will fix at 200 revolutions per minute. If the dynamo, and consequently the engine, were smaller, this latter would run faster, and *vice versa*. We must bear in mind to stipulate for a heavily-balanced fly-wheel on each side of the crank; the necessary tanks to connect with pipes to the jacket on the cylinder, in order that the water in them will circulate, and so keep the cylinder cool; and all other details and spare parts.

The selection of the dynamo is an easy matter, since all that it is necessary to state, is that it must be shunt-wound; that it must give 40 ampères at 135 volts, at (say) a speed of 1200 revolutions per minute with a double-flanged pulley of 12 inches in diameter, since the fly-wheel on the engine will be six feet in diameter, and will make 200 revolutions per minute. An endless double leather or link belt ⅜-inch thick will be required, and the dynamo must be driven with the belt tight on the lower side.

The accumulator shelves or racks, if there is plenty of room, can be about 30 inches from the floor, in a single tier ranged round the walls. They must be in a separate room from the dynamo, and between these rooms there must be no communication. For this reason, it is best to select one from which we have to come out into the open air before we can enter the dynamo-house, so that there will be little risk of the spray, which the cells give off on completion of the charge, being blown in upon the dynamo, and so damaging the insulation. The ventilation of an accumulator-house by means of a skylight or roof-ventilator is less likely to allow the fumes to attack young trees or shrubs, than an open window. The floor of this room must be such that it can be washed down into a drain.

The foundations may be made of Portland-cement concrete, that of the engine 4 inches, and that of the dynamo 10 inches above the level of the floor, which may be either boarded over, or covered with oil-cloth or tiles.

The switchboard will be conveniently fixed about 18 inches from the wall, near the dynamo. The distance between the engine-shaft centre and that of the dynamo-shaft centre should, considering the high speed, be about 14 feet, and as the fly-wheels of the engine will require pits, the floor must be channelled out, so that the belt may not touch as it runs.

The water-circulating tanks we will fix near the engine-cylinder, in the corner, on a stillage a few inches from the ground, and we will supply them with a ball-cock to ensure the water-level being maintained. It is a serious matter for the cooling of the cylinder to cease, as would be the case if the level of the water dropped below the uppermost pipe, because the cylinder might be damaged through the piston "siezing" in it.

The mains, consisting of 2000 megohm insulation of vulcanized rubber, covered with lead, we will lay about 12 inches under the turf in rough tarred-wood casing with a central rib, similar to that used inside buildings, only less ornamental. All cinders must be carefully removed before screwing on the cover, as they would set up an electrolytic action and eat away the lead. These lead-covered cables would be connected to the switchboard, and the lead covering must be stripped off for about 6 inches, before they make contact thereto. Also immediately on their entering the house at the other end, they should be similarly treated, and led into two single-pole cut-outs. By this arrangement the surface-leakage along the lead on to the cut-outs will be stayed, and these cables can at any time be isolated for testing.

From these two cut-outs in the house, we will proceed with ordinary 600 megohm wire to the first **distributing board**, and thence to the other distributing boards, at a limit of 5-ampère circuits,—that is to say, each circuit must not have more than 8 lamps, each taking ·6 of an ampère.

Great care must be displayed in **the selection and arrangement of the lamps and fittings.** Commencing at the entrance,—if there is a lamp there,—great care must be taken, not only to place it entirely out of reach of actual water, but as far as possible to protect it from moist air, the latter being quite sufficient to cause the lamp to fail.

The carriage-drive lamp, if such should exist, we will supply by two lead-covered wires in a trough (similar to those previously described), running from two cut-outs in the house-cellar right up into the globe of the pedestal, which should be weatherproof, remembering to strip back the lead covering at each end for a few inches; and the switch we will place in the vestibule, so that a servant can light the lamp without leaving the house.

The lamps in the hall should be provided with so-called "corridor" switches, one being in the hall, and the other on the landing upstairs, or in the master's bedroom, both these switches being of special construction, and for which special but simple wiring is necessary, so that the lamps can be lit or extinguished by either switch. Thus, at dusk, the lamps can be lit by a servant in the hall, and at night extinguished by the master after he has retired to his bedroom, without putting the switch in the hall out of action at all for the next day.

Corridor switches can also be fitted in bedrooms, one at the door immediately on entering, and one at the head of the bed.

In the drawing-room, connection wall-sockets are often required for portable lamps, and to prepare for all future demands it is as well to run a pair of wires right round the room on the wainscot or in the dado rail, to which to connect them without causing any damage.

With a little thought, **switches and their wires** can generally be entirely hidden if desired, and to effect this, a point to remember is that the wires can possibly be led to any point on the far side of the wall on which they are to be fixed. Given sufficient time and money, it is generally possible to wire a house without any of the casings being visible in the reception rooms, or the best bedrooms.

In selecting the course for the wires, water-pipes, which may burst, and troughs or baths, which may overflow, should be carefully avoided. Special care should be exercised in false roofs. In stables and along damp cellar walls, it is desirable to "pack out" the casing by small pieces of wood, driven in and standing about one inch from the wall, and to varnish the casings with shellac inside and out, back and front.

When the work is completed, we may find that **the exhaust from the engine** makes a distressing noise. In such a case, it must be turned into a pit, dug just outside the engine-room, about four feet square and deep, lined with fire-clay bricks, drained, flagged at the bottom, and filled with cobbles the size of a man's fist, or gradually diminishing in size towards the top to the size of walnuts. If the exhaust-pipe is turned into the bottom of this horizontally, and drilled with $\frac{1}{4}$-inch holes all over its length in the pit, and the pit is roofed over, allowing a wind-space of about 12 inches between the roof and the ground-level, the trouble will be cured.

Sometimes **the belt runs against the flange of the dynamo-pulley,** persistently showing that it is out of line; this may perhaps be cured by simply "slewing" the dynamo round a little on its rails.

After testing the insulation, and all being found in order, we proceed to

charge the cells. To take every precaution against a mistake, we connect to the two wires leading from the engine-room for such charging, two small wires, and with their other ends we connect each of them to a piece of lead about six inches long by two inches wide, or to a piece of lead tubing; each of these two pieces we fasten to a piece of wood, in order to keep them apart, and then stand them in a small jar of acid, having previously placed in the circuit of this small cell, which we have made, a 100-volt lamp. We now run the dynamo full speed to give 135 volts as shown by the voltmeter, and reduce this with our shunt-resistance to 100 volts, at which pressure the lamp will allow only ·6 of an ampère to pass through the jar. After a few minutes one of the lead plates will have turned brown, which will indicate that that wire is connected to the positive terminal of the dynamo. The positive terminal of each cell is painted red, and the negative black, and as the cells are being erected, red should be joined to black. When this has been done, the positive wire from the dynamo must be joined to the positive terminal of one of the end cells, and, having poured in the acid, we are ready for charging.

The following should be the course adopted in charging. Before starting the engine we connect the voltmeter across the accumulators, with a small switch on the board for that purpose, and find they register 80 volts. This is very low, and the cells will never register so low again, but being new they are absolutely discharged. We now start the engine, and run the dynamo at full speed, when, as we know, it will give 135 volts, but this we reduce with the resistance to 90 volts, or ten more than the cells give. We then close the cell-switch, and see by the ammeter that a few ampères are going into the cells. Thus we have established the current without causing a sudden rush, and can now easily increase it as desired. To do so, we merely watch the ammeter, and as we reduce the resistance in the shunt-circuit and so raise the E.M.F., we see the current rise to 40 ampères, which we must not exceed. Indeed, with these new cells it will be best to charge at 30 for about ten hours, when we will proceed at the 40 ampères for the remaining twenty hours, for which time, at least, we expect to have to charge on this first occasion. Many sets of cells have been ruined by being insufficiently charged in the first instance. The charging must be continued until all the cells boil, and turn quite opaque with "gasing".

When these cells are charged, they will give at least 2 volts each, so that a "many-way" switch is provided, connecting (by a separate wire to each cell) the first few cells in the circuit; standing at the board when the charging is out, we can, therefore, connect the cells to the voltmeter as previously, and connect only such a number of cells to the circuit as will give 100 volts. Since, then, some of the cells are not working all the time, they do not require as much charging, and so these are provided also with a many-way switch. The operator with these two switches can charge and discharge as many of the cells as he wishes.

If the belt happens to come off the dynamo, or the engine to stop, the dynamo will have no power, and the cells will send a current into it, since they are connected to it—and a powerful current too, as, when we commenced to charge them, they had an E.M.F. of 80 volts. Of course the dynamo cut-outs would be fused, but to prevent this frequently happening, **an automatic accumulator-switch** is used, which closes the dynamo-circuit when it is of a superior E.M.F. to that of the cells, and opens it immediately it is of an inferior voltage to theirs.

All the wires from the cell-room, leading to the many-way switches, must be fitted with **cut-outs in the cell-room**, close against the cells, to prevent the slightest possibility of a short circuit.

We will notice, before leaving the engine-room, that **printed instructions** have been framed and hung up to instruct any attendant hereafter appointed in the management of the plant; and that a copper or brass oil-can has been provided for the dynamo, so that it may not be attracted by the magnetism (as it would be if of ordinary tinned iron) and drawn into the dynamo, perhaps doing £20 worth of damage, and that dynamo-lubricating and engine-lubricating oil has been provided, with a filter for cleaning it for use again; the second time, however, it should be used for the engine. And on entering the house, we will see that similar notices are hung in the servants' hall, to be there for many years after perhaps the present servants have left, pointing out that on any damp arising in the house from any of the causes described, the wires to that portion of the house are to be cut off at the distributing-board, by removing the two fuse-wires, until the damage is repaired.

SECTION XIII.—LIGHTING.

PART II.—GAS.

BY

HENRY CLAY.

CHAPTER I.

METERS.

All gas-meters bear a government stamp, and at times the stamp of the gas company, whether it be public or private. The government stamp, affixed after testing and payment of the fee according to the size, is a guarantee of honesty (for the time being), but such guarantee is little regarded by gas companies; some curious instances might easily be given to show that these meters soon have to be removed on receipt of notice. Gas companies favour a particular class and kind of meter, and it generally happens that only those meters likely to develop a "galloping consumption" are supplied by them. Meters that do not show a gradual increase of consumption, or that do not at least register the full amount of gas consumed, are those that are soon condemned, the nominal fault being either "not registering correctly", or "escape at index", "index-glass broken", "seal broken", or "requires testing", although any of these faults may exist in favoured kinds for years, if the consumer makes no complaint.

Meters are either "wet" or "dry", the former being the favourites; but there are great differences in wet meters, some costing little more than half the price of others. Some have cast-iron cases, the more costly having tinned-iron cases. There are some slight differences in regard to the formation of the drum and overflow arrangements, and one meter is known as the "unvarying water-line", which seems to condemn all other kinds as unreliable, and consumers are naturally led to think that such a meter is about the best of its kind for them, whereas the facts are all the other way, for it errs always on the side of the gas company and against the consumer after it has been in use some time. The dry meters register more correctly, but, if the pressure in the street-mains is increased, they may pass a little more than indicated. At times they stop registering altogether, but continue to supply gas, notwithstanding the index-pointers being fixed.

In **wet meters** the quantity of gas is measured in the drum, which is suspended on an axle in the water, as shown in fig. 640, and the revolutions of the drum are transmitted from the axle to the index by toothed gearing

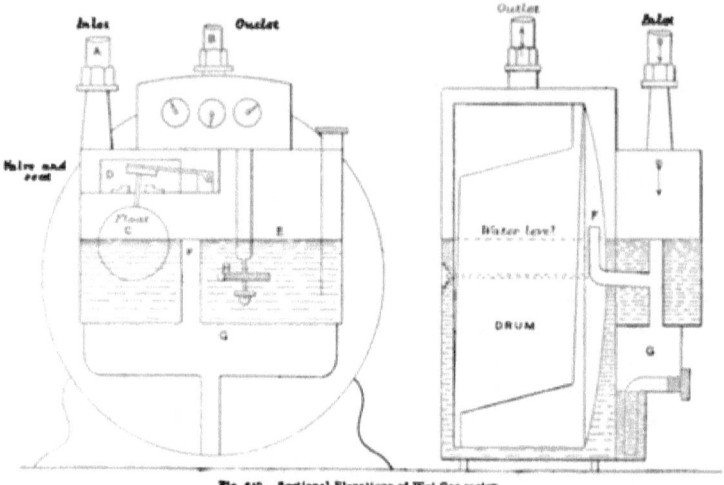

Fig. 640.—Sectional Elevations of Wet Gas meter.
A, inlet; B, outlet; C, float; D, valve and seat; E, water level; F, supply pipe to drum; G, drainer box; H, pinion and wheel operating the dial.

according to the capacity of the drum. Wet meters require to be fixed perfectly level on account of the revolving drum, spindle, water-level, and float, although at times they have to be packed owing to the drum being slightly wrong. The drums of wet meters are made of pure sheet tin (white metal). As the water falls the drum revolves more and more slowly, until at length the quantity passed is not sufficient to supply the demand, when all the lights will begin to jump, and, unless the meter is attended to, the gas will stop altogether. This is owing to the float C gradually sinking, until at last the valve seats itself at D, and stops the supply of gas from passing through the meter. It will be seen from this that the wet meter is constructed to stop the supply when the water is low, or, in other words, as soon as it begins to supply more gas than is registered.

In **dry meters** (fig. 641) the quantity of gas passing through is measured in the bellows, which open and close alternately. As the gas enters the bellows on one side of the partitions, the bellows on the other side are closed, and the measured quantity of gas forced out of them and through a slide-valve operated by levers and cranks in connection with the bellows and index. The slide-valves prevent the return of the measured gas to the bellows, and open and close alternately with the supply-valves to the bellows. The leather of the bellows sometimes becomes clogged with gas liquor, or stiff with the moisture, but the commonest cause of failure is an escape at the index, owing to the stuffing-boxes being worn out. When the bellows become fixed, the index is also fixed, as the movement of the bellows caused by the pressure of the gas is communicated to all parts, and, although the whole of the movement may be stopped, there may be a small supply of gas through leaky stuffing-boxes and valves. Dry meters will work when fixed unevenly, but they are much better when fixed level. Dry meters are not affected by frost.

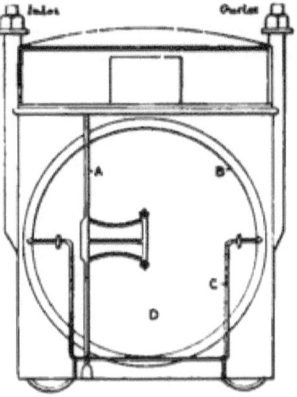

Fig. 641.—Sectional Elevation of Dry Gas Meter

A, Spindle in communication with dial; B, the leather of the bellows when closed; C, brass guide and support to bellows; D, tin plate.

The dry meter supplies **the gas at the outlet** in the same condition as it receives it. This can hardly be said of the wet meter, as the water in it becomes anything but sweet; and it is just possible that, if a little more attention were paid to the periodical renewal of the water in wet meters, the air of gas-lighted rooms would be considerably improved.

The water in wet meters may freeze during frost, and a slight thickness of ice across the surface and on the sides is sufficient to hold the drum fast. The manner of applying heat to thaw the ice within a wet meter will depend upon the position of the meter. In some cases hot water can be used internally and externally; in others, flannels dipped in hot water; and, where water cannot be used, hot sand may be placed on and around the meter; and if sand is not available, use bags of salt. Hot bricks may be placed against the meter, but are not recommended, as at least one death has resulted from their use. Bricks are easily heated until they are red-hot, and when placed against the front of meters having tinned-iron cases, the soldered seams are melted.

Where **the gas-mains** are below the reach of frost, the incoming gas usually keeps the water in the meter from freezing, if it is not exposed to a frosty current of air, so that with deep mains, and the meter protected by a wood casing filled with saw-dust or hair-felt, it will be safe from frost. Where the gas-mains are above the ground, they should be well protected by being wrapped with hair-felt or other good non-conductor, to prevent the gas from being chilled to such an extent as to absorb the heat from the water in the meter as it passes through, thus causing the water to solidify during the time it is in use. In such cases the ice will be broken in pieces as soon as formed, but the accumulation will soon be sufficient to prevent the drum revolving and supplying gas. A layer of glycerine on the surface of the water in a wet meter is good to prevent loss of heat from the surface when the meter is at rest.

TABLE XXXIX.
GAS-METERS—THEIR INLETS AND OUTLETS, NUMBER OF LIGHTS, &c.

Name of Meter.	Size of Inlets and Outlets	Number of Lights allowed.	Stamping Fee.
2-light	$\frac{1}{2}$ inch	5	6d.
3-light	$\frac{3}{4}$,,	8	6d.
5-light	$\frac{3}{4}$,,	12	6d.
10-light	1 ,,	22	1s.
15-light	1 ,,	30	1s.
20-light	$1\frac{1}{4}$,,	40	1s.
30-light	$1\frac{1}{2}$,,	60	1s.
50-light	$1\frac{1}{2}$,,	100	2s.
60-light	$1\frac{1}{2}$,,	120	2s.
80-light	$1\frac{3}{4}$,,	160	3s.
100-light	2 ,,	200	3s.
150-light	3 ,,	300	5s.

The sizes of gas-meters are known by the number of lights, as given in the first column of the Table XXXIX., which are standard numbers for all kinds of gas-meters regardless of their shape or class. The sizes of the inlets and outlets given in the second column correspond with the quantity of gas passing through meters of the several capacities. The figures in the third column give the maximum number of lights that may be supplied by any of the meters without interfering with their correct working. All meters from 10 lights upwards are capable of supplying at least double the number of jets, a 10-light meter being able to supply 22 separate jets, and a 100-light meter 200 separate jets. The figures in the first and third columns are enforced in order that the sizes of meters and pipes shall be based upon the number of jets.

For economical working, the gas-meter should always be as small as possible, in order to thoroughly break up the pressure from the mains when the meter is not being worked up to its total capacity, and without causing the lights to fluctuate. At the same time, care must be taken in places where a good steady light is necessary, to have the meter large enough to allow of a slight decrease of pressure in the street-main, without interfering with the supply of gas passing through, and in such cases the total number of lights should not exceed the numbers given in the third column of the table, and special main gas-governors and governor-burners should be used.

Up to 100 lights **the connections to the meters** are by means of brass unions, and from 100 lights upwards the connection will be by bolted flanges.

The stamping fees vary with the size of meter, being the same for all makes. Meters bearing the stamp may be fixed anywhere, notwithstanding any rules of gas companies to the contrary; but it is unwise to compel any company to accept meters they object to, as the company will take care that such meters always have something wrong with them, and will make them very costly. When meters are condemned, the seals are broken and the meters are sent to the repairing shop, and are afterwards tested and restamped.

The durability of meters varies according to their size. Small meters generally work from eight to twelve years before they need repairs, and last from fifteen to twenty years. The larger sizes (from 50 to 200 lights) last much longer, but much depends upon the position in which they are fixed, and the amount of work they have to do.

CHAPTER II.

GAS-PIPES AND FITTINGS, &c.

Lead pipe is the most suitable for conveying gas in houses. It can be readily bent and fixed in all positions. The joinings can be soundly made, even when numerous connections come close together, and alterations or additions are easily effected. Lead pipes do not corrode, and no dust is given off to choke up the fittings or burners. There are, however, several disadvantages in the use of lead gas-pipes, for, if badly laid, they sag, forming loops or festoons between the fastenings, and these loops become water-logged owing to the condensation of vapour, and cause the lights to jump and ultimately go out. In houses subject to vibration lead pipes break, usually about the staircases, and they are also liable to be pierced with nails when hanging pictures or ornaments.

Lead gas-pipes are made from inferior lead, and when buried in the plaster the lime frequently affects the lead to such an extent as to cause it to crack every few inches.

Composition pipes are not so much used now as formerly. They are harder and stiffer than lead, and not so liable to be affected by lime. They require a little more skill to join neatly, but the joints are equally as good and sound as the joints on lead pipes.

Iron pipes are most suitable for workshops, and all other places where the pipes are exposed, or have to be suspended. They are sometimes used in houses, and give no further trouble after being once properly fixed, except from rust and dirt.

The service-pipe from the street-main to about 6 inches inside the house is usually of iron with ordinary screwed joints. Galvanized-iron pipes are used in some places, and lead in others. Both lead and plain iron service-pipes should be surrounded with a loose wood casing, which, in the case of the iron, is afterwards filled with a composition of pitch.

Outside the premises **a stop-cock** is fixed for the use of the company's men, and another is always fixed as soon as the main enters the building. The connection from this stop-cock to the meter is usually of lead, the same size as the service-pipe and tap, and is soldered to the unions provided by means of a copper-bit or spirit blow-pipe, using resin as a flux.

The outlet-pipe from the meter will be the same size as the inlet, and will be continued—in the case of houses—past the centre of the first floor, when it can be reduced to suit the number of lights supplied by the branches. The supply-pipe to the second floor will also be of less bore, and the same again applies to the third floor. With a 10-light or 15-light meter, the supply to the first floor will be 1 inch, to the second $\frac{3}{4}$ inch or $\frac{1}{2}$ inch, and to the third $\frac{1}{2}$ inch or $\frac{3}{8}$ inch. Where there are four lights to be supplied on the first and second floors, a $\frac{1}{2}$-inch branch should be run past the first two, and afterwards continued with $\frac{3}{8}$-inch pipe. The same four lights on the third floor would only require $\frac{3}{8}$-inch pipes throughout, owing to the pressure increasing the higher the pipes are carried. In good work, $\frac{1}{4}$-inch pipes are seldom used on account of their weakness, leakages being frequent, and because a very small drop of water is sufficient to cause the lights to jump. If these pipes are used, they should never be fixed horizontally on this account, and in some places they are not allowed at all.

In fixing the gas-pipes care must be taken to have a continuous fall back to the meter, in order to prevent the lodgment of the condensed vapour in any part of the pipes. If sufficient care is taken in arranging the runs, there will be no necessity to fix syphons; indeed syphons are usually a sign of unskilled labour. They were at one time pretty numerous, but are now rarely met with, except where alterations or additions have been made.

When the main service-pipes are laid in private ground between the house and the street-main, and the distance is too great to allow of a fall to the street-main, one or more **syphon-boxes** will be necessary. All such boxes are provided with taps to drain off the water collected in them. When the gas-mains are continued outside the house to supply the out-buildings, syphons will also be required according to the position and length of the pipes. A short length of iron pipe with the end capped up, or a piece of lead pipe with a tap at the end, is often used instead of a syphon-box, but requires to be emptied oftener and more regularly than the latter.

Short lengths of pipe having blank ends are termed **syphons** (the boxes used as receivers being termed syphon-boxes), not because they are syphons, but because their predecessors were inverted syphons,—that is, a pipe bent in the form of the letter U,—the water forming a trap and preventing the escape of gas, so long as the water draining from the pipes equalled or exceeded that evaporating from the exposed end of the syphon. In many cases the quantity evaporated exceeded the supply, with the result that serious explosions occurred, and the use of the syphon was prohibited; but the name still remains, and although syphon-pipes and syphon-boxes are terms often met with, it must be understood that they have no connection with the syphon proper, as they are merely pipes or boxes fixed to receive and retain the condensed water.

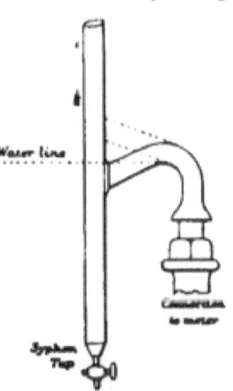

Fig. 642.—Connection of Internal Gas-main to Dry Meter.

In the case of wet meters up to 50 lights, the pipes must be laid to fall to the meter, **the condensed water** helping to keep the meter charged. Where a dry meter is fixed, the water must on no account be allowed to drain into it. The main supply-pipe in the building should be carried down below the meter, and the outlet-pipe from the meter branched into it in such a way as to stop the gas before the water can overflow into the meter when the syphon is full; this can be done, as shown in fig 642, by bending down the end of the outlet-pipe so as to bring the branch-joint below the underside of the bent pipe, thus forming a trap, which will indicate by the fluctuations of the gas-lights that the syphon is full. If connected as shown by the dotted lines, the water can drain into the meter.

The **taps to gas-fittings and supply-pipes** are usually of the plain plug type, having a hole straight through the plug, a quarter-turn either way turning the gas on or off. In some fittings, such as hall-lamps, the plugs of the taps have only one hole in the side communicating with the interior of the hollow plug, and a quarter-turn serves to turn it on or off, but the plug must be turned backwards and forwards through the same quarter space, instead of the two half spaces in the taps of chandeliers, brackets, and pendants. Main gas-taps for lead or iron pipes are plug taps of the ordinary kind, having a quarter-turn.

Gas-governors are fittings arranged to open and close the gas-tap on the main supply according to the number of lights in use. The face of the indicator may be arranged for any number of lights, and by turning the pointer to the number of lights in use, the gas-tap is opened or closed to correspond, thus allowing only the amount of gas consumed by the number of lights in use to pass through the meter. The great advantage of an arrangement of this kind is that, no matter how high the pressure may be in the street-main, it is completely broken up when the governor is properly regulated, and no loss is caused to the consumer by excessive pressure. It is useless to try and regulate the gas-tap without an automatic arrangement of this kind, for the regulation of the tap must be in proportion to the number of lights in use, and must be altered as lights are added or put out. These regulators may be fixed in any convenient position.

Bruce, Peebles, & Co.'s mercurial gas-governor for mains (fig. 643) is not so complicated in its parts as most others. There is no possibility of water lodging in it in such a way as to interfere with its working. It is claimed for this governor that it comes into action at a lower pressure than most others, and as all governors supplied are tested before leaving the works, they will be properly regulated. It may, however, be pointed out that the weighting arrangements are not as sensitive as they might be, and to regulate the bell by means of these weights requires much more patience and skill than the ordinary gas-fitter is likely to possess.

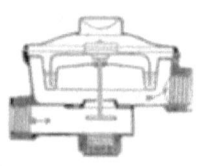

Fig. 643.—Section of Bruce, Peebles, & Co.'s Mercurial Gas-governor

In the **"Wenham"** gas-governor (fig. 644) the difference in the weighting arrangements will be at once apparent. The principle of the governor is the same, including a bell, mercury seal, and two valves, instead of the one in the former. In addition to the improved weighting arrangements (which consist of a lever with movable weight and screw), the interior of the governor is arranged to prevent it opening and closing too rapidly. In use this governor will be

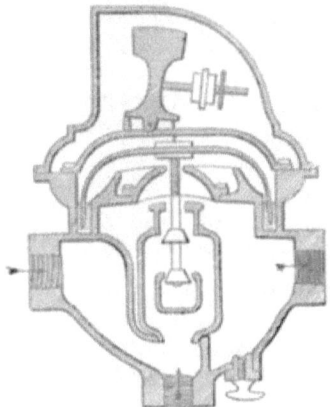

Fig. 644.—Section of the Wenham Gas-governor.

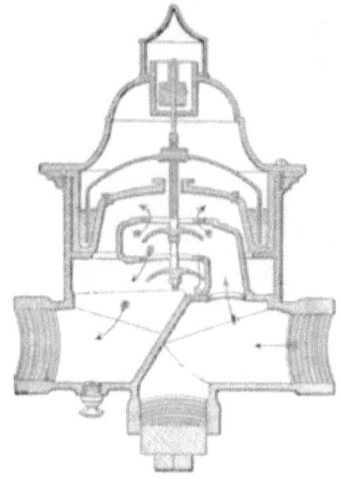

Fig. 645.—Section of the Stott Gas-governor

much more steady in operation, and can always be relied upon to correct the flow of gas without causing the lights to fluctuate.

The "Shaw" gas-governor is in principle the same as the "Wenham", with the exception of the weighting arrangements. There is little difference between the "Stott" and the "Shaw" gas-governors. In principle and weighting they are alike. In construction they differ slightly, the ways in the "Stott" (fig. 645) being larger and clearer than in the "Shaw" (fig. 646). The construction of the "Shaw" allows it be taken to pieces and cleaned much more readily than the "Stott". In regard to actual gas-governing capabilities, they may be set down as exactly alike, there being no difference in principle or weighting; but

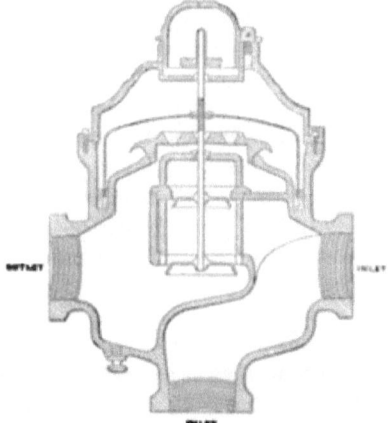

Fig. 646.—Section of the Shaw Gas-governor.

when compared with the "Wenham", both the "Stott" and the "Shaw" fail, owing to the faulty system of applying the weights.

The "needle" governor burner (fig. 647) is a gas-burner constructed to pass a given quantity of gas, and to thoroughly break up the pressure, thus maintaining an uniform consumption under varying pressures. These burners effect a considerable saving of gas, and it is impossible to have a smoky flame, as the whole of the gas is consumed, and the maximum amount of illumination (with an open-flame burner) is therefore attained. The "automatic" governor (fig. 648), made by the same firm, is designed to control the quantity of gas supplied to gas-fires and stoves. By means of the regulating screw underneath, the quantity of gas delivered can be easily adjusted to suit the capacity of the stove or fire. It is a simple and efficient governor.

Fig. 648.—Bruce, Peebles, & Co.'s "Automatic" Gas-governor for Stoves and Fires.

By-pass taps are used for keeping small jets continually burning in such positions that the whole of the lights in connection with them will be properly lit by simply turning the gas on to the full. They are also used in churches and theatres, where the lights are often turned down to prevent overheating as well as unnecessary consumption. By-pass taps may be regulated to give a mere glimmer of light or up to a quarter of the full supply. Small by-pass taps are necessary where the fittings are in such positions as to be practically inaccessible for lighting by hand, and in many places they save a large amount of time, as all that is necessary where they are provided is to turn on the supply.

The connection of gas-pipes to fittings, including stoves, gas-fires, and cooking-ranges, should be by brass unions. In some cases the unions will form part of the gas-taps, and in others they will be complete in themselves. The unions should be those known as ground unions, requiring no packing materials. They should also be used when connecting pipes of different metals, such as iron to lead, and lead to copper.

Blocks or pateras are used for the purpose of fixing brackets and pendants. The block is secured to a fixing in the wall or ceiling, and the fitting is afterwards secured to the block. In the commoner class of fittings the blocks are from 4 to 8 inches in diameter, and $\frac{3}{4}$ to 1 inch thick, the edge moulded or carved. In the better class of fittings the blocks must be made to match the back of the fitting, the shapes of which have been considerably improved during recent years.

CHAPTER III.

GAS-BURNERS.

The ordinary Bray's burner is the one in most general use. It is in no sense a governor, although it is arranged to check the pressure, and requires to be selected according to the amount of light required, and the pressure of the gas-supply. Only a governor-burner can be fixed without regard to the pressure of the supply, and as the "Bray" is not a governor, a selection must be made. Three kinds are shown in section in fig. 649. The commonest form A has a canvas screen fixed inside the body, and the "special" burner has an enamel plug in addition to the canvas screen. The hole through the enamel plug is graduated to suit the various pressures, and the holes or slits in the burner heads. To get the full amount of illumination obtainable, the size of the burner must be decided upon after testing the various sizes. No. 3 and No. 4 union-jet burners give good results in most cases. All the "specials" are good when selected to suit the varying requirements in each particular case.

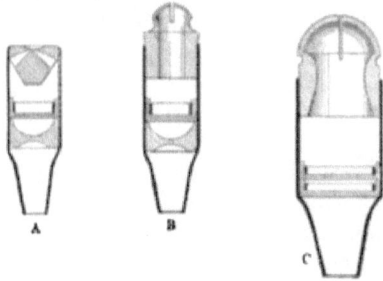

Fig. 649.—Sections of three Bray's Gas-burners

A, burner with union jets; B, special slit-union burner; C, standard slit-union burner (made in sizes from 10 to 80 candle-power)

The consumption of gas with any particular size of burner will vary with the pressure as well as the quality of the gas supplied. This also applies to the candle-power of all burners; the tabulated results of tests, giving illuminating power, consumption, and pressure, are of little value, except as showing the varying efficiencies at the particular time and place of testing. The quality of the gas supplied, as well as the pressure, varies so much that it would be unfair to select any series of tests and use them to decide the relative merits of the various burners. To set down the tabulated results of tests made for the rival manufacturers of gas appliances would be very misleading, as in most cases such tests are made at the pressure most suitable for the particular burner. In a test made to ascertain the effect of increased pressure upon the quantity of gas consumed, it was found that, with a small burner at two-tenths pressure, there was a consumption of 1·890 cubic feet per hour; and, increasing the pressure by tenths, it was found that, at ten-tenths, there was a consumption of 4·800 cubic feet per hour, the average increase of consumption for every tenth of pressure being 0·3 cubic feet per hour. Here the quality of the gas and the illuminating power of the burner are neglected, but it will be seen that the consumption of gas increases or decreases according to the pressure. In order to compare fairly the tests of rival burners, it is necessary to know not only the pressure and the consumption of gas in cubic feet per hour, as well as the candle-power, but also the quality of the gas. The standard illuminating power of the gas supplied in the various cities and towns varies considerably, and, besides this, there is always a slight variation in quality even in the same locality.

The **albo-carbon burner** was designed to achieve the carburation of coal-gas by the admixture of hydro-carbon vapour. It consists of a receptacle or generator for the "albo-carbon", this material being vaporized and mixed with the gas before the flame is reached. It is claimed that, by using an 8-light albo-carbon cluster of the kind suitable for shop windows, the lighting power is over $8\frac{1}{2}$ standard candles for every cubic foot of gas burned, as against 3 candles per cubic foot when gas is burned in the ordinary way. To charge the generator, unscrew the cap of the feeding-screw provided, and fill the vessel with pieces of the albo-carbon, taking care not to injure the screw, and also to properly screw up the cap to prevent leakage. When the gas is first lighted, the albo-carbon in the generator will be solid, and some few minutes must elapse before the heat from the flame is sufficient to volatilize it, and thus cause it to issue from the burners with the coal-gas, which is thoroughly carburetted by the admixture of the hydro-carbon vapour. Care must be taken not to move or shake the generator whilst warm, as the albo-carbon is then in a liquid state, and will block up the tubes and burners. It is a cheap and effective system, well recognized and appreciated by the public, who have adopted it where a brilliant and steady illuminant is required at a moderate cost, but it is somewhat dirty in use.

Argand burners are a great improvement on the ordinary burner, and give a steady light of considerable power. They were first invented by Aimé Argand, who was born at Geneva in 1755, and were of course designed for oil-lamps. The principle has since been successfully applied to gas-burners. It consists essentially of an annular cylinder; the gas enters the annular space and escapes through a series of small holes in the upper edge of the cylinder, forming a circular flame, to which the air has access on both sides. A chimney is necessary to create a draught and steady the flame, and to obtain the greatest illumination. A section of Sugg's argand burner is shown in fig. 650, which, although a little more complicated, answers to the foregoing description.

Regenerative gas-burners, which are all more or less indebted to Siemens's original patent, now form a very important class of fittings, the central idea being to raise the air and gas to a high temperature (by means of the products of combustion) on their way to the actual burner.

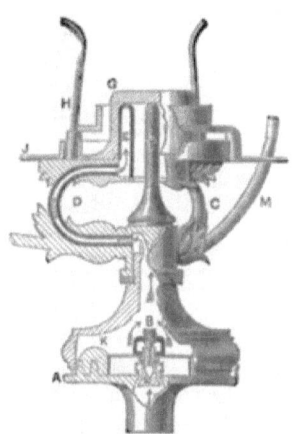

Fig. 650.—Section of Sugg's Argand Burner.

A, bottom plate; B, screw regulating supply to burner from gasholder; E; C and D, two, out of three, tubes through which the gas is conducted to combustion chamber, N; G, air-cone, regulating delivery of air to outside of flame; H, one of the wires which hold chimney; J, wire supporting globe; K, leathern gas-holder to which is attached valve, L, controlling admission of gas; N, support for shade.

Fig. 651.—Elevation of Wenham Lamp.

The Wenham gas-lamps are of this kind. The ordinary Wenham lamp, illustrated in fig. 651, consists of a burner inclosed in a glass cup. The air-supply to the burner is drawn through the outer casing, and passes away along with the products of combustion through the inner tube back into the apartment. During the passage of the air to the burner, it receives heat from the inner tube, and is redelivered into the apartment in a highly-heated state. These lamps more nearly approach a gas-stove than any other fitting I know, and are, on this account, unsuitable for household lighting purposes, unless a special outlet flue is provided for the products of combustion, as shown in fig. 652. The light itself is good and steady when properly regulated and governed, but this is of course the difficulty.

These lamps require to be cleaned occasionally, the inner tube at times being choked with soot, especially when the spreader of the burner is faulty, or the pressure and quality of gas unsuitable. When these lamps explode in lighting, the soot is loosened and the inner tube partially cleared, the soot floating about and settling on the various articles in the room. When by-pass taps are provided, there will be no explosion on lighting, but there is more probability of the inner tube being blocked with soot. They are most suitable for halls, corridors, and other places where there is a free current of air.

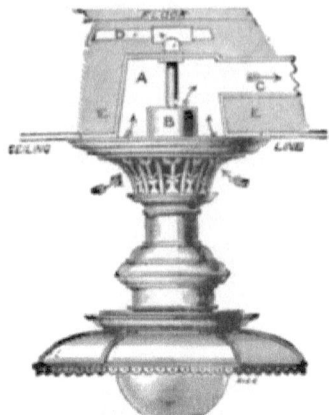

Fig. 652.—Wenham Ventilating Lamp.

A, iron ceiling-box; B, outlet for vitiated air, &c., from lamp; C, sheet-iron outlet shaft to flue; D, gas-supply pipe; E, non-conducting material around box and shaft.

A vertical section through the Wenham burner is given in fig. 653. A is the actual burner of cylindrical form, held in position by the ring B; if this becomes loose, gas will leak inside and prevent the gas burning white. The button D, if bent or made untrue, will also distort the flame. The gas enters at L and is heated in its passage through the spaces F F. The heated air passes through the gratings C C and G G, and so impinges on both sides of the ring flame, which is shown by the dotted lines M M. It will be noticed that the flame is the lowest part of the lamp, and consequently no shadow is cast downwards.

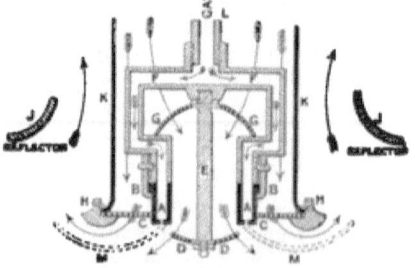

Fig. 653.—Vertical Section through the Wenham Burner.

A, perforated porcelain cylindrical burner; B, ring holding the burner in position; C, perforated disc; D, perforated button; E, rod supporting the button; F, way for gas; G, perforated dome; H, ring for supporting globe, secured by a bayonet catch; J, reflector; K, external cylinder; L, gas-supply pipe; M, the flame.

The **incandescent gas-light** possesses great advantages over the older methods of illumination, and even over the electric light. It is claimed that (1) it saves half the gas required for ordinary burners, whilst giving at least three times the light; (2) the reduced consumption of gas results in less carbonic acid and less heat being generated than with ordinary burners; (3) the burner being on the bunsen principle, and being so constructed as to ensure perfect combustion, is almost smokeless. There are now several rival incandescent burners, but the Welsbach is still the best.[1]

The "C" Welsbach burner gives a light of about 60 candle-powers, with a consumption of 3½ cubic feet of gas per hour. The ordinary open-flame gas-burner gives about 15 candle-powers, with 7 cubic feet of gas per hour. The illumination from the incandescent gas-burner is therefore about four times greater, whilst using only one-half the gas.

With the common type of gas-burner there is practically no expense in first cost and maintenance, but from the figures given it will be seen that the payment for the gas passing through them is altogether out of proportion to the quantity of light obtained. It would require four ordinary gas-burners, each consuming 7 cubic feet per hour, a total of 28 cubic feet per hour, to equal the illumination from one "C" incandescent burner, consuming 3½ cubic feet per hour. The extra cost of the incandescent gas-burners, together with the renewals of mantles, chimneys, and globes,[2] will be saved in from one to two years according to the amount of gas consumed. When compared with the best regenerative systems of gas-lighting, the incandescent light does not show so large a saving. In order to get a light of 280 candle-powers, with gas at three shillings per thousand cubic feet—

With plain burners the cost would be about 4d. per hour.
  „ argand „ „ „ 3d. „
  „ regenerative „ „ „ 1d. „
  „ incandescent "C" burners „ $\frac{9}{16}d$. „

[1] Experiments by Prof. W. Wedding, of Berlin, on the relative value of five different incandescent burners, were published in the *Journal für Gasbeleuchtung* for 1895, and the following table contains an abstract of the results. The table is interesting not only for purposes of comparison, but also because it shows the gradual deterioration of even the best mantles.

TABLE XXXIX A.
COMPARISON OF INCANDESCENT GAS-BURNERS.

Name of Burner.	Consumption of Gas per Hour.		Illuminating Power.	
	When first lighted.	After 222 hours' use.	When first lighted.	After 222 hours' use.
	cub. ft.	cub. ft.	candles.	candles.
Auer (Welsbach)	3·81	3·85	57·3	35·4
Trendal	3·67	3·85	20·1	13·2
Bense	3·74	3·58	23·6	6·4
Stobwasser	3·60	3·92	31·6	4·0
Kramme	3·60	3·58	22·7	9·6

Further particulars of the experiments will be found in *The Proceedings of the Institution of Civil Engineers*, Vol. CXXI, Part III.—ED.

[2] The number of breakages is very largely reduced when the burners are suspended from springs and the gas admitted by flexible tubes, as in the A.V.I.L. apparatus; this is now adapted for brackets as well as for pendants.—ED.

The cost of the incandescent light is therefore about one-half that of the regenerative, and only one-eighth that of the plain burners.

The following is taken from the report of Professor Carlton Lambert, M.A., F.R.A.S., dated July 27, 1894:—" The Welsbach light is nearly seven times as efficient in illuminating effect as ordinary gas-burners, more than four times as efficient as an "argand", and more than twice as efficient as a regenerative lamp. From theoretical considerations, it is evident that the quantity of carbonic acid actually produced *by each foot of gas burned* must be the same with the "Welsbach" as with the "Bray", or any other burner with which the consumption is perfect. The vitiation of the air by the "Welsbach" light is almost negligible, being between one-fifth and one-sixth part of that produced by ordinary burners *giving the same amount of illumination.* The heat evolved by "Welsbach" burners is a little less than one-seventh part of that emitted by ordinary gas-burners *giving the same illumination.*" I have italicized portions of this quotation, lest they should be overlooked. The illuminating power of the "Welsbach" burners is excessive, and, in house-lighting, a large proportion is wasted. One "Welsbach" giving 60 candle-powers should be sufficient for a medium-sized room requiring four No. 5 Bray burners. In practice it will be found that, owing to the intensity of the light, heavily-muffled and tinted globes and shades must be used, necessitating the fixing of (perhaps) three "Welsbach" burners instead of one. For this reason the comparisons in the last two sentences of the report are somewhat misleading.

The special "Lancet" Commission reported in January, 1895, much more clearly, the four most important "conclusions" being as follows:—

"1. The burner (with or without the mantle) is not prejudicial to health.

"2. The *heat and carbonic acid* produced by the combustion is [sic] one-half *those of an ordinary gas-burner,* though the light given is more than three times as great.

"3. The *heat and carbonic acid* produced is [sic] *less* even than those of *the average oil-lamp.* . . .

"5. There is not a trace of any poisonous gases—such as carbon-monoxide or acetylene—in the products of combustion."

It is clear from these reports that the incandescent system of gas-lighting is to be highly recommended on account of its illuminating power, small consumption of gas, and consequent slight vitiation of the atmosphere.

INCANDESCENT GAS-BURNERS. 295

There are five descriptions of Welsbach incandescent burners issued for domestic lighting:—

1. The "C" burner.
2.   „ „ with governor.
3.   „ „ „ by-pass.
4.   „ „ „ governor and by-pass.
5.   „ "S" „ „ by-pass.

The construction of the burner, and the method of taking it to pieces for cleaning or regulating, and also for the refitting of mantles, will be understood from figs. 654, 655, 656, and 657.

The "C" burner, complete with mantle and chimney, is shown in fig. 654, about one-half the real size, and the several parts are shown in fig. 655. This

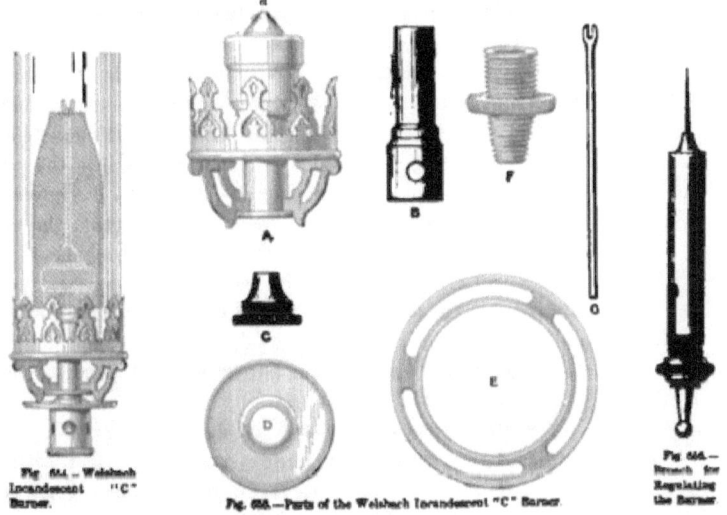

Fig. 654.—Welsbach Incandescent "C" Burner.

Fig. 655.—Parts of the Welsbach Incandescent "C" Burner.

Fig. 656.—Broach for Regulating the Burner.

burner is regulated to consume $3\frac{1}{2}$ cubic feet of gas per hour, at a pressure of one inch. If, however, for local requirements, it is necessary to regulate the burner to suit a higher or lower pressure, instructions to that effect should be given when ordering. The regulating of the burner may, however, be done by means of the "broach", illustrated in fig. 656. If the holes in the nipple c, fig. 655, are too large, they can be burnished down by the round end of the broach, until the holes are sufficiently small to pass just the amount of gas required. Should the holes be too small, then the needle in the same broach is used to increase their size. Only a very slight enlargement of the holes is necessary to make them of the required size. In fig. 655 A is the gallery, on the top of which is a small hole a, into which fits the central mantle-rod G. It also supports the chimney; and the globe-ring E, which carries the globe, slips over and rests on the outside of the gallery. The part B is the bunsen tube, and C the nipple, which, when screwed together, form the complete bunsen burner. D shows a plate which slips over the bunsen as far as the shoulder, and prevents the lighting back so frequent in all bunsen burners. F is a brass nose-bit or "adapter" with male threads, the larger end made to screw into the bottom of the nipple under the burner, and the tapered end to screw into the burner-socket on the gas-fitting.

The "C" burner with governor is in all respects the same as the ordinary "C" burner, with the addition of a small governor underneath, and, in consequence of this, the air-holes of the bunsen burner have to be made somewhat larger. This "C" burner with governor is recommended in all instances where the pressure of gas varies greatly. There is no doubt that the duration of the mantle and chimney is greatly shortened by extreme variation in the pressure of the gas, and a great saving will be effected if a governor is used, as it keeps the bunsen flame always steady, the consumption of gas is always the same, and the breaking of the mantles and chimneys is greatly reduced. The governor is so adjusted that the necessary amount of gas to bring the mantle to full incandescence is obtained with $\frac{10}{16}$ths pressure, and however the pressure may increase, the burner will not consume more than $3\frac{1}{2}$ cubic feet of gas per hour.

Fig. 657 — Welsbach Incandescent "C" Burner with By-pass.

The "C" burner with by-pass is shown in fig. 657. The feature of this burner is a lever arrangement, either with or without chains, which, on being turned up after opening the gas-tap, allows the burner to be lighted as would be the ordinary "C" burner, but if the gas-tap is left open, and the lever of the by-pass turned down, then the bunsen flame is extinguished, but at the same time a small jet is kept burning inside the mantle. By this arrangement the relighting of the bunsen burner can be effected by simply raising the by-pass lever again. It is a very simple arrangement, and strongly recommended.

Great care should be taken with burners of this kind, as the small brass tube inside the bunsen burner must always be kept in its proper position. By taking off the gallery from the bunsen tube, it will be seen that inside it is a small brass tube. When the gallery is replaced, care must be taken that the lighting-back plate stands with its curved edge upwards, and that the three bottom brass points of the gallery come right down on to it. The object of this is to ensure that the little tube enters the socket inside the gallery. If the tube does not enter the socket, then the by-pass will not work properly, and the gas will bubble on the gauze of the bunsen burner, make an unpleasant noise, and create an objectionable smell by escaping. Every burner is manufactured so that this small tube fits exactly into the inner hole of the by-pass; but sometimes it gets knocked out of its position, and then it must be bent so that it fits correctly again into the hole referred to, in the inside brass top of the gallery.

By the employment of the by-pass burner a great economy in the use of gas may be obtained, as people using it would soon get into the habit of turning down the gas when it is not required, feeling that there would be no trouble in relighting it when necessary. The by-pass jet consumes about 2 cubic feet of gas in 24 hours, so the consumption of gas need not be taken into account, as, if it burned for the whole year with gas at 3s. per thousand, it would not consume more than about 2s. worth. In bedrooms the by-pass jet answers the purpose of a night-light, as it gives just sufficient light to mark its position but not more.

The "C" burner with by-pass and governor, is the ordinary "C" burner with the additions of by-pass tap and governor as described above.

The "S" burner with by-pass is made on the same principle as the "C" burner with by-pass. It gives approximately only 30 to 35 candles, and for house-purposes this is often quite sufficient. This burner is regulated to consume 2¼ cubic feet of gas per hour, at 1 inch pressure. In comparison with an ordinary Bray burner, it consumes only one-half the quantity of gas, whilst giving double the amount of light, one-half the quantity of heat, and one-half the quantity of carbonic acid.

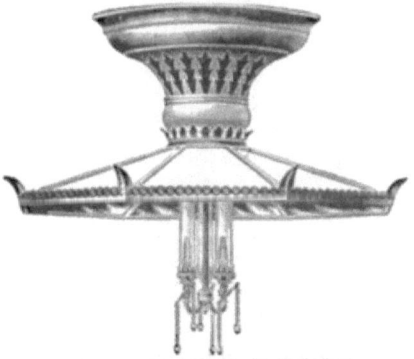

Fig. 656.—Three-light Incandescent Ventilating Lamp.

As **the quantity of heat** given off by all burners (including the incandescent) is the same per cubic foot of gas consumed, the incandescent (consuming as it does only one-half the amount required by a No. 5 Bray burner) will only give off half the amount of heat, and will therefore be less efficient when used for the purpose of extracting air for ventilation. Where there are clusters of incandescent lights, however, the total amount of heat given off will be a valuable aid to ventilation, as the up-currents through the chimneys will carry the whole of the heat and carbonic acid away through the flues provided, as well as induce a current of air from the apartment. Two ventilating lamps are shown in figs. 658 and 659; the smaller (fig. 658) carries away the products of combustion from the burners, and ventilates the room by withdrawing air at the level of the ceiling. In the larger lamp (fig. 659) there are no openings at the ceiling-level; the air must enter from beneath the glass shade, making it impossible to thoroughly change the air in a moderate-sized room. All the heat and products of combustion pass away through the casing to the flue provided above the fitting.

Fig. 659 — Large Incandescent Ventilating Lamp.

The principle adopted is the same as in the "Wenham" lamp, as both lamps draw the supply of air (necessary for the complete combustion of the gas passing through the burners) from the apartment, and pass it away, by means of an inside tube, to the flue above. Both kinds of lamps are also adapted to induce a current of air to pass through the perforations at the ceiling-level, in the direction of and through the same flue as the heated air and products of combustion. The "Wenham" lamp is much superior to the incandescent for the purpose of creating and maintaining a current of air in one direction from the apartment, owing to the consumption of gas, and the heat given off being greater. Provision must also be made for a larger volume of air to pass through the casings of the "Wenham", thus ensuring a more rapid and thorough ventilation of the apartment.

With most ventilating lamps, **reflectors of various shapes and materials**—for fixing internally or externally and sometimes both—are required to reflect the light downwards, owing to the lights themselves being kept as near as possible to the ceilings. Opal reflectors do not darken the ceilings above them so much as metal or silvered glass reflectors, and are therefore much better. It is not customary to use reflectors with other classes of house-fittings, except in the case of billiard-table lights and reading-lamps, and with both these fittings the cardboard shades are used more to protect the eyesight, owing to the flame being much lower than when used for ordinary lighting purposes. Such shades are usually green outside and white inside, the white acting as a reflector, but not to the same extent as the materials usually employed. Silvered copper and glass are the best reflectors, but both appear to be fast dying out, porcelain-enamelled iron being now used when subjected to great heat, and opal glass in other cases.

The burners used for oil-gas are ordinary "Brays", but instead of using Nos. 3, 4, or 5 as with coal-gas, burners numbered 0000, 000, 00 are required. Although these sizes appear very small in comparison with the ordinary coal-gas burners, they will be found to give a light at least equal to the coal-gas burners, owing to the greater illuminating power of the oil-gas. It is also found that this gas gives better results than coal-gas, when it is consumed in the Welsbach incandescent burners.

CHAPTER IV.

GAS-LEAKAGES AND EXPLOSIONS.

One of the most frequent causes of explosion is the faulty gas-tap. All these taps have a small pin or stop inserted in the plug to prevent the taps from being turned past the centre, in such a way as to turn out the light, and afterwards allow of the gas escaping. These pins or stops often work loose and drop out, and when this occurs it is a very easy matter to turn the tap too far when turning off the gas, thus turning it on again; the person can merely guess as to the position of the plug by the thumb-piece, which may also, and in many cases is, twisted out of its true position, so that it is next to impossible to tell when the plug of the tap is in its proper place. Where such pins or stops are missing, it is advisable to have them renewed at once.

The **plugs of gas-taps** often work slack, and a slight leakage of gas occurs from them. When these slack plugs are left without attention, the nuts and washers generally fall off, leaving the plug loose. When the screw-nut hole enters the hollow of the plug, a gas-escape occurs which must receive attention, but in most cases these holes do not enter the gas-way, and as no leakage occurs if the plug is pushed in every time it is used, they are often left without attention until some person accidently leaves the plug out of its position, when it is quite possible for an explosion to result, as the leakage will soon fill the apartment.

Brackets and pendants often come loose from their fixing, and in both cases, unless the gas-fitter is called in, there will be a strong leakage. When fittings come loose, the strain on the gas-pipes soon breaks them, and in either case the leakage is in a dangerous place—i.e. in the floor or walls.

Water-slide pendants and chandeliers are the most fruitful source of explosion. In both cases, two or three table-spoonfuls of water will be sufficient to keep them gas-tight, when they are pushed up as high as they will go, but if they are drawn down to their full extent, the loss by evaporation of the same quantity of water will soon cause a very large leakage of gas, which may fill the apartment, and, if unnoticed by a person entering with a light, may cause a disastrous explosion. When fitting new globes to chandeliers, great care should be taken to have the weights balanced to suit, as it is possible by merely changing the globes to overbalance the weights; this will not matter much so long as the tubes are full of water, but will be dangerous where, owing to fires being kept in the room, evaporation is rapid. Another source of danger arises from the rapid deterioration of the brass chains of all chandeliers. If one of the heavy weights breaks its chain and falls, the body of the chandelier being relieved will slide down to its full extent, when there is every probability of a leakage, if it has not been attended to recently. When these weights fall they usually alarm the household, but there are many places where the fall of these weights would not be noticed, even in the daytime.

Many explosions have occurred through **allowing water to accumulate in the pipes**, not in sufficient quantity to interfere with the supply, but enough to make a slight rattle in the pipes, when the mice, hearing the movement of the water, may gnaw the pipe away until the gas issues.

It will be obvious then, that the whole of the gas fittings and pipes require to be periodically overhauled by a competent gas-fitter, the cost of which may be less than the price of the gas escaping from the numerous taps and joints of the fittings.

When a leakage of gas occurs, the first thing to do is to give the apartment air by opening both windows and doors. Then go down to the gas-meter, and turn off the main-tap above the meter provided for the purpose. Next send for a gas-fitter, and if he lights a candle to search for the escape send him away and get another. A good gas-fitter will not risk an explosion for the sake of a few minutes' time, but will cut the pipes and test them, two cuts being all that are usually required, even in a large residence, to locate the position of the leakage.

Section XIV.
GAS-PRODUCING APPARATUS
FOR THE
ILLUMINATION OF COUNTRY HOUSES

BY

J. MURRAY SOMERVILLE
CHIEF ENGINEER'S STAFF, SOUTH METROPOLITAN GAS COMPANY

SECTION XIV.
GAS-PRODUCING APPARATUS FOR THE ILLUMINATION OF COUNTRY HOUSES.

CHAPTER I.
COAL-GAS.

Four kinds of illuminating gas are now in practical use, the *first* being that obtained from the distillation of coal, the *second* from the retorting of oil, the *third* from the chemical action of water upon carbide of calcium, and the *fourth* from the evaporation of light volatile hydrocarbon liquids. Each of these gases has its peculiar advantages and disadvantages, which, with the processes of manufacture, will now be briefly described.

A suitable coal for carbonization can be obtained in small quantities of 10 tons, costing 14s. per ton delivered in London. A ton of this material can be made to produce from 10,000 to 10,500 cubic feet of gas in a small gas-works of capacity sufficient for 50 lights; the cost of this gas, even on such a small scale of working, should not exceed two shillings per 1000 cubic feet.

The buildings and plant for the production of coal-gas consist of a retort-house, coal-store, condensing-apparatus, tar-tank, purifiers, meter, and gasholder.

The retort-house and coal-store should be weather-proof, and well-ventilated, but free from draughts. There is invariably a ventilating roof to retort-houses, as shown in Plate XXIV. The apparatus contained in the building consists of the retort-benches, hydraulic main, the coal, stoking and charging implements, pressure-gauges, valve-keys, and, generally, all tools or apparatus which would deteriorate or perish on exposure to the weather. The retort-bench contains two cast-iron retorts, 7 feet long, of a ⌒ section about 16 inches wide and 12 inches high, and cast with a closed or "blank" end at one extremity, and an open or "flanged" end at the other. A furnace or fire-grate is placed under each retort, so that they are independent of one another as regards the making of gas. The exit-flue from each furnace runs into a chimney common to both. Retorts and furnaces are contained in a square body of brickwork of fire-clay material, held together by iron straps or buckstaves to prevent cracks, which would otherwise be caused by the expansion and contraction of the brickwork. An arch is turned over the retorts to relieve them of any pressure from the structure, and also to allow the hot gases from the furnace to have full play around the retort.

It is very seldom that a gas-retort is fused, but the continual exposure to atmospheric influences causes an oxidation or "perishing" of the iron. With a badly-constructed furnace, this deteriorating action is rapid; with one having a considerable depth of fuel, the period of life is greatly increased. The minimum life of a retort having a thickness of metal of 2 inches, is about six months; the maximum, with gaseous and careful firing, may exceed eighteen months.

On to the outside flange of the open and front end of the retort,—which is buried about 2 inches into the front wall,—a casting, called the "**mouthpiece**", is attached. This mouthpiece is of a similar section to the retort, and is about 9 inches long. On the upper side is cast a socket, whose inside diameter is an inch to an inch-and-a-half greater than the upright ascension-pipe which it receives. A gas-tight joint is made between the socket and pipe by a "rust-joint" composed of sal-ammoniac and iron filings. At the extremity of the mouthpiece, there is a hinged or slotted-bar arrangement for carrying the door of the retort, which by a mechanical contrivance is pressed upon the smooth surface of the mouthpiece, or upon a "buttering" of moist lime and sand, for the purpose of "sealing" the retort after it has been charged.

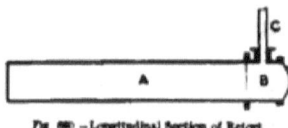

Fig 680 —Longitudinal Section of Retort.
A. retort. B. mouthpiece; C. ascension pipe fitting into socket of mouthpiece

The gas given off by the destructive distillation of the coal comes forward to the mouthpiece, and passes upwards and through the "ascension-pipe", by which it is conveyed to the "hydraulic main" placed on the top of the retort-bench. The hydraulic main is generally supported by small brick piers, or cast-iron brackets, at the height of three or four feet above the level of the bench, in order to allow a free circulation of air to protect it from the reflected heat of the retort-bench.

Hydraulic mains are generally of a half-round section, **having about 6** inches **depth of water at the bottom.** The "ascension-pipe" is continued or carried over, **either by a half-round pipe, or by a short** horizontal length of circular pipe, **to**

the hydraulic main, and is connected to it in such a manner that the end of the former is covered by one or two inches of the water in the latter. The gas consequently has an upward and downward motion before reaching the hydraulic main. This part of the apparatus is provided with a cover-plate and a gas-outlet, and can be arranged to receive the gas from several retorts. The water covering the ends of the pipes entering the trough, technically known as "dip-pipes", acts as a "seal", preventing any gas returning to the retorts, and consequently rendering these independent of each other, though they are connected with the same hydraulic main.

The piping from the outlet of the hydraulic main to the next portion of the plant, known as the "condenser", is generally taken round the inside of the retort-house, for the purpose of reducing the temperature of the gas before entering the condensing-apparatus.

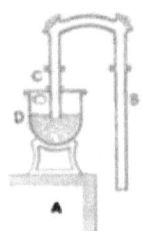

Fig 681.—Section of Hydraulic Main.
A, retort hatch; B, ascension pipe; C, dip-pipe; D, hydraulic main.

The remaining apparatus and plant required in the manufacture are generally well coated with paint or tar, and placed in the open; so before proceeding with their description, a short account of the work in the retort-house will be given. Upon the entrance of the stoker to withdraw a charge of coal which has been subjected to a sufficient period of distillation, his first attention should be directed to the pressure-gauges, as the indications of the pressure of every portion of the plant serve to show the condition of the whole works. Any excess exhibited demands instant attention and release before the retorts are re-charged. The fire in the furnace will require disturbance in order to throw down the ash and refuse and brighten the fuel, before re-stoking. Having cleared the fire-grate, the retort-lid or door is taken away, a rake having a "crook" at the end is inserted into the retort, and the solid carbonaceous substance therein withdrawn. A portion of this hot fuel is fed into the furnace, and the remainder quenched into "coke". Any tarry and pitchy particles that may adhere to the orifice of the ascension-pipe are removed, and a short iron-rod pipe-cleaner, or "augur", run up into the pipe. If a "luted" lid or door is used, the luting is now done, and the fresh coal shovelled into the retort; the lid is then placed in its position and screwed home or otherwise fastened, and gas-making again commences.

Some coal deposits tarry substances in the "ascension"-pipe, and thus causes considerable trouble. The ascension-pipe should be provided with caps, which can be opened for the insertion of a clearing-rod. It is sometimes necessary to light a few sticks of wood, previously soaked in naphtha, underneath the pipe, after having opened the cap or joint on the top of the pipe. In most cases, however, the constant use of the clearing-rod will prove an ample means of providing clear ascension-pipes. A quantity—the greater quantity, in fact—of tar is deposited in the hydraulic main, and is allowed to run to the tar-well through the same pipe which conveys the gas to the condensers. It is therefore of some importance to keep this residual product in a fluid state during its circulation in the retort-house.

At certain intervals, attention must be given to **the quantity of water or "liquor" in the hydraulic main**; the tar should never be allowed to rise so as to cover the ends of the dip-pipes, for obvious reasons. The liquor deposited from the gas is frequently insufficient to furnish the requisite supply of "seal", and water must therefore be added as required. A gauge, connected to the top and bottom of the hydraulic main, will serve to indicate the condition of the seal.

It is usual to place a **cast-iron trough underneath the furnace** for heating the retort-benches, and to run water into it, so that any ash or small pieces of fuel which may escape through the bars of the grate are quenched, and the steam thus produced rises up into the fire, cooling the fire-bars on its way.

In retort-benches of this description furnaces of a very old-fashioned and elementary character are generally placed, requiring attention every two or three hours. As the charge of a retort requires six hours before it is thoroughly carbonized, it would obviously be a **saving of time** if all the work could be done in the retort-house when the charges must be withdrawn, that is to say, four times per diem; the time so occupied would not exceed one hour for each charge. This method of arranging the manufacture is easily accomplished, with very little extra cost, by having a furnace of suitable capacity only requiring disturbance four times in every twenty-four hours.

In **the storage of coal**, it is important to "work out" every heap placed in store. Coal allowed to stand deteriorates more or less according to the length of exposure to the atmosphere.

With the exception of **charging and discharging the purifiers** at various periods according to the quantity of gas made, the foregoing description of the duties attending gas-manufacture will very closely represent the daily routine.

The condenser, which is the first portion of the outdoor plant to receive the crude gas from the retort-house, aims at the removal of tarry matters and condensable vapours from the gas. In small works, this is accomplished by allowing the gas to slowly ascend and descend a number of somewhat larger pipes, placed in an upright position over a trough or long tank, which is partitioned in such a manner as to ensure the circulation of the gas. The gas leaves a suitable and efficient condensing-apparatus, in a clean and clear condition to the naked eye. The trough which receives the lower ends of the upright condenser-pipes serves to hold the tar, which separates from the gas during its circuit owing to the diminished rate of progress, and the reduction of the temperature of the gas upon exposure to the atmosphere.

The tar is eventually run into a storage-tank, usually placed underground on account of the objectionable odour of the tar. The tar must be allowed to stand in the tank at least three days in order to "settle", *i.e.* to allow any intermingled liquid to rise and float upon the surface; consequently, in emptying the tank the tar should be withdrawn from the lower level. A gallon of this tar, at a high temperature,—somewhat under boiling-point,—mixed with one-and-a-half to two pints of boiled linseed-oil, makes a very excellent covering-material for iron and wood, not only on account of its preserving qualities, but also from its appearance, the mixture being capable of producing a very lasting lustre. It should undoubtedly be used in gas-works, owing to its resistance to the action of fumes.

A quantity of liquid, known as "ammoniacal liquor", is found on the top of the tar. In works of small size this "liquor", and the lime used in the purification of the gas, are generally known as the two "nuisances". The lime can be used beneficially on the land, but the "liquor" must be disposed of in any convenient and suitable way.

The diminutive size of the works is also a reason for the omission of the apparatus known as the "scrubber", in which gas is subjected to a prolonged contact with moistened surfaces, for the purpose of absorbing the ammonia. That office is therefore performed by the purifiers.

There are two **purifiers, or purifying boxes**. They are square in shape, and resemble an ordinary box with a cover or lid. The bottom and sides are generally made of cast-iron plates; the box is about 3 feet deep, and has a recess 6 to 8 inches deep and about 6 inches wide running around outside the top of the side plates, technically called "the lute". This recess is filled with water, and the lid or cover is so arranged that, when in position, its sides sink into this water, which thus forms a "seal" or "lute" and confines the gas within the vessel. The cover is usually made of wrought-iron. Within the box, attachments are made for horizontally supporting the "grids", which carry the

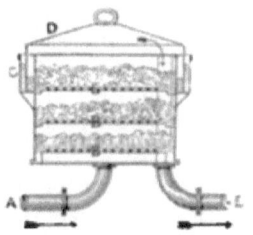

Fig 441—Section of Purifier.
A, Inlet-pipe; B B B, grids carrying the purifying material; C, lute; D, cover; E, outlet-pipe.

material used in purification. The grids are made of wooden strips about 4 feet long, 2 inches deep, and three-quarters of an inch in thickness, spaced about half an inch from one another by packing pieces, and each grid carries a layer of lime or oxide of iron 6 inches deep. There are usually three tiers of grids in each box. The gas-inlet to the purifier is attached to the bottom plates, and the gas is forced up through the interstices of the purifying medium, and thus brought into contact with absorbents of its pernicious ingredients. The outlet for the gas is cast at one corner of the box formed by the joint of two side plates.

The object of purifying the gas is to remove from it two deleterious components, namely, sulphuretted hydrogen and carbonic acid. The first compound should be completely removed, owing to its poisonous and destructive properties, its presence being shown very rapidly by the gray tinge which it gives to all lead paint, and the brown tinge it produces on all gilding, plating, and silver work. The presence of carbonic acid in the gas renders the illumination less brilliant. The appearance of both impurities can be rapidly and easily ascertained, and should ensure the immediate discharge of the purifying material from a "box". The means of turning the stream of gas from one purifier to another is invariably provided in gas-works. It is known as a "by-pass", and by its use the gas is passed through a clean purifier, while the other, which has become "foul", is being changed. Soft gray lime proves the best purifying agent, and should not be slaked till within twenty-four hours of using.

The purified gas is led to the meter, and from thence to the gasholder. The meter is an instrument designed for the measurement of the volume of gas made, but it is not a necessary adjunct to a gas-works, as a gasholder will act as an approximate "tell-tale" of the quantity of gas made and consumed.

A gasholder for the capacity of works under our notice would have an outside diameter of 12 to 16 feet. The structure consists (as shown in Plate XXIV.) of an outer tank about 8 feet deep, made of wrought-iron sheeting, and having a ring of angle iron, or member of some similar section, around the top edge. Attached to this tank are the columns and guides, which rise in a vertical position rather more than 8 feet above the rail of the tank. Inside this tank is placed an inverted tank of slightly-diminished diameter, which is restrained from any side or circular movement by roller guides fixed on the top and bottom, which engage with guides connected to the inside of the columns. The only movements allowed to this "holder" are consequently those of rising and falling. When filled with gas the holder rises, owing to the effect of the pressure exerted on the top or "crown". The same pressure is felt upon the water placed in the tank to form a "seal", and is counteracted by lifting a quantity of the liquid outside in the annular space between the holder and the inside of the tank, which is only exposed to the pressure of the atmosphere. The difference in height between the outside and inside levels of water in the gasholder-tank gives what is technically known as the "gas-pressure" at that point; in other words, gas-pressure is the height of a column of water in air maintained by the pressure of gas. It is an invariable object to reduce the pressure of gasholders; the wrought-iron sheeting of the holder is therefore of very light material. The inlet and outlet gas-pipes to the holder are carried up through the bottom plate of the tank above the level of the water.

A "governor" is placed on the outlet-pipe from the gasholder, for the purpose of modifying any variations in the pressure of the gas due to the rising and falling of the holder, or depressions given to it by gusts of wind. Such differences in pressure are instantly felt where this instrument is not used, causing the flames in burners to "jump" or vary in size rapidly, instead of burning with a steady flame as is the case under a constant pressure of gas. The position of the "governor" is frequently close to the point of consumption. This apparatus[1] is of small bulk, and can be fitted inside the house without difficulty or detriment, and is so designed that, with either five or fifty burners in action, the pressure of gas supplied to them remains constant.

Coal-gas is without doubt the most amenable and well-conditioned gas for purposes of illumination. It is a mixture of compounds of hydro-carbons, other carbon bodies, and hydrogen, very stable in composition, and of light density. It is an excellent heat-agent, and easily distributed. It possesses one unfortunate characteristic, viz. that of depositing naphthalene crystals in the outdoor service-pipes, but this never becomes a source of trouble in small works, and rarely occurs when cannel-coal or shale is used. Preference should be given, however, to Durham coal or Yorkshire Silkstone, on the score of cheapness, and freedom from benzine constituents in the composition of the gas. Further, where the illumination is effected by means of the excellent ventilating regenerative burners,—such as the Wenham[2] and Vertmarche,—the illuminating power of a cubic foot of cannel gas is very slightly above that of gas produced from common coal.

[1] See figs. 643-646, pp. 256-287, Vol. II.—ED.   [2] See figs. 651-653, pp. 291-292, Vol. II.—ED.

CHAPTER II.

OIL-GAS.

There are two distinct systems of producing illuminating gas by the retorting of oil. The direct cracking and gasifying in iron retorts is the plan in general use, and fractional distillation in peculiar iron vessels is a recent development.

The first of the above methods is exemplified in the well-known **Pintsch's oil-gas process**, which is also extensively used in the lighting of railway-carriages. The carbonizing plant for the supply of 150 lights is shown in fig. 663, and consists of a bench of four retorts in two tiers, arranged so that

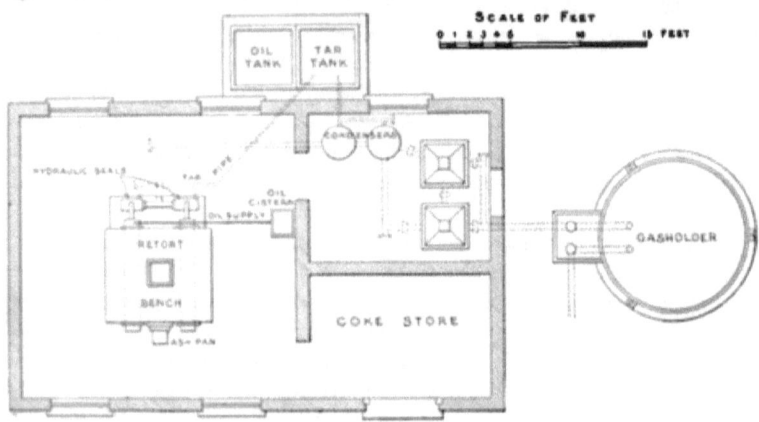

Fig. 663.—Ground-floor Plan of Works for Pintsch's Oil-gas Process.

each upper retort is connected to the lower one beneath it, by a short pipe, and the two ovens, thus coupled, form one complete retort for the vaporization of oil. Only one "couple" of retorts is used at a time, the other being a stand-by for use during repairs, &c.

The retorts are from six-and-a-half to seven feet in length, and are supported by cross-walls over a furnace of ordinary type. They vary in shape, but are mostly of ⌒ section, 10 inches wide and 8 inches in height. They are heated by direct firing from the furnace below, the lower retort having a temperature somewhat in excess of the upper one, for the specific purpose of completing the gasification of the oil. Each "couple" has a separate and distinct hydraulic seal.

OIL-GAS MANUFACTURE.

The oil to be vaporized is run into the top retort through a tube having two bends, forming a **natural seal,** and technically known as a "gooseneck" seal, as shown in fig. 664. It will readily be seen that a sufficient head of oil **must** be decanted into the seal to cause the liquid to flow into the retort against the pressure from within. As the resulting **gas has to force its** way through a hydraulic seal, condenser, and purifiers, and into the gasholder, this pressure is frequently sufficient to support a **column of oil** seven or eight inches high. Under such conditions it would be necessary to obtain a level of liquid some seven inches higher in the longer limb of the U-tube, to ensure the free passage of the oil into the retort.

The oil flowing into the upper retort is conducted for about three feet of its length in a wrought-iron trough, permitting it to receive a little heat before coming into contact with the cast-iron retort, lest the latter should fracture on account of the sudden extremes of temperature. In this retort the more volatile con-

Fig. 664.—"Gooseneck" Seal for Running Oil into Retort.

stituents of the oil are immediately converted into vapour, and pass with the heavier hydrocarbons into the lower retort. The latter performs two functions. The remaining heavy hydrocarbons are gasified, and the whole of the vapours and crude gases are subjected to the reflected heat of this retort, which tends to resolve the mixture into less complex bodies and compounds of more stable character. Such action is technically known as "fixing", or superheating. So that in explaining the use of the two retorts, we may say generally, that the upper retort "cracks" or vaporizes the oil, and the lower retort fixes the vapours into illuminating hydrocarbon gases.

The manufacture of oil-gas is carried on with **very little labour or supervision.** Having adjusted the flow of oil to the temperature of the retorts, the production of gas, in many plants on this system, continues without appreciable intermission for a period of six months. The life of the retorts is rather less than those used for the distillation of coal. The scale or carbon deposit, left in oil retorts, is more difficult to remove, and at times appears to be fused into the iron as a form of slag. The "life" of a retort, however, may safely be put at six months, and each retort can be detached and replaced without difficulty and at small expense.

The gas, as it leaves the lower retort, **passes through a** hydraulic seal of small dimensions, kept cool by the passage of a continuous supply of water through the vessel. A large quantity of the heavy hydrocarbons is deposited at this point as tar, the liquid being carried by an overflow-pipe to the tar-well. From the seal, the gas passes to the condenser. The hydraulic seal and condenser perform similar offices to those used in coal-gas plants, the one to prevent "back pressure" into the retort, and the other to remove the last particles of condensible compounds.

The condensers in oil-gas plants are known as "atmospheric condensers", being so designed as to allow the surrounding air to circulate freely between the upright pipes, or annular casing, down which the gas is conducted. By this means a constant uptake of cool air robs the gas of its high temperature, and, together with the diminished flow generally provided for in this apparatus, causes a quantity of unstable compounds to separate out as heavy oil or light tar. In large installations this liquid is frequently run off separately, as it forms a very excellent liquid fuel, and can be used in a "spray". The heavier tar is invariably burnt in the furnace under the retorts for economizing fuel. The total quantity of tar thus obtained generally amounts to about one-third of the oil used.

From the condenser the gas is passed through purifiers to remove sulphur compounds, and finally into the gasholder. For the purpose of lighting railway-carriages, the gas is compressed in iron cylinders to a pressure of from 12 to 15 atmospheres, but for domestic use a pressure of 6 inches proves very suitable.

The quality of the gas is very high, being four times as rich as coal-gas made from ordinary or common coal; its illuminating power ranges from 55 to 65 candles. In consequence of this property, the amount of gas consumed is considerably diminished. Flat-flame burners for this gas are designed for a consumption of one-half to three-quarters of a cubic foot per hour, giving an illuminating power of 10 to 12 candles. Regenerative burners for the consumption of this gas are now in vogue, increasing its efficiency as an illuminant, and an agreeable soft light of 40 candle-powers can be procured by a consumption of 2 feet per hour.

The amount of oil which can be properly treated by each couple of retorts varies with the temperature to which they are raised, a general quantity being $2\frac{1}{2}$ gallons per hour. The yield of purified gas, taking a fair average, is 80 cubic feet per gallon of oil. The supply of gas for 150 lights in winter would necessitate the carbonization of eleven gallons per diem. In practice, however, the plant would be operated twice a week, and the gasholder designed to store three or four days' consumption. A very good oil for the purpose of gasification can be procured at a cost of $2\frac{1}{4}d.$ to $2\frac{1}{2}d.$ per gallon. The cost of this oil-gas will be about 3s. per 1000 cubic feet.

Heating and cooking stoves have been designed for the special use of oil-gas and other gases very rich in hydrocarbons, and give satisfaction in every way equal to stoves burning coal-gas.

Owing to the slight attention required, the freedom from nuisance, the diminutive size, and the low cost of the plant, this oil-gas system should obtain consideration. There are two items, which, being satisfactorily dealt with, should favour its adoption. First, the supply of fuel for the furnace, and secondly, the carriage of the oil.

A more modern method of obtaining illuminating gas from oil is the invention of Mr. Young, and is known as the Peebles process. In this process the oil is subjected to a very ingenious system of fractional distillation. A cylindrical bottle-shaped iron retort, about seven feet long, having a diameter of two feet in the body, gradually reduced to one foot at the mouth, has a spherical blank or closed end. The retort is laid with a slight fall towards the closed end, and is heated to a red or dull red heat by a furnace, or by waste gases from other sources. Oil is allowed to run into the retort in such a stream that it is vaporized before it has travelled the entire length. The lighter constituents are vaporized in the neck of the vessel; the remaining portions (consisting of the more dense hydrocarbons) flow towards the end, and thus receive a prolonged exposure to the heat of the retort, which is necessary to convert them into vapour and gases. Owing to the considerable diameter of the retort, the vapours are exposed to a large amount of surface-contact and heat-reflection, and the effect of these is felt for a longer period by those heavier hydrocarbon gases generated at the end of the retort. The object of this arrangement is recognized when we consider that the lighter hydrocarbons are stable compounds of high illuminating power, and the character of the heavier hydocarbons undergoes a change of an improving nature upon subjection to a superheating or "fixing" action. It is observed, therefore, that in this process the exposure of the vapours from the more volatile constituents of the oil to the reflected heat of the retort is considerably less in comparison.

The gas thus produced passes through an oil seal to prevent "back pressure" in the retort, travels along a pipe conveying the oil to the retort, through a "washer" or "scrubber" (where the gas, in small divided streams, is brought into intimate contact with the oil), and finally into a purifier and gasholder. In its passage through seal, oil-pipe, and "scrubber", those constituents of the gas which are not of a stable character are dissolved by the oil, and by this agency returned to the retort to undergo another process of "cracking" and "fixing". In this manner, gas of a permanent description is produced. A great advantage of this system is that no tar is produced.

The practical results of plants on this system show that the whole of the oil cannot be converted into gas. A small quantity is lost in the formation of a hard brilliantly-crystalline coke, which has a great value in metallurgical manipulations, and which can always be sold for over £1 per ton. The volume of gas obtained from a gallon of oil treated by this process lies between 85 and 88 cubic feet, having an illuminating power of 60 candles. The deposit of coke is about 2 lbs. per gallon of oil.

The conditions under which the oil is treated in this system of gasification are very scientific, and are attended with an increased yield of gas. The adoption of this process, however, should greatly depend upon the supply of fuel for heating the retort. Oil-gas by this arrangement is obtained at a cost of 2s. 7d. per 1000 cubic feet.

CHAPTER III.

ACETYLENE.

In the year 1892, while making some experiments with an electric furnace, Mr. Thomas Leopold Wilson discovered that powdered coal and lime, mixed together in equal parts, fused under the influence of very high temperatures, and eventually entered into combination, forming the compound known as **carbide of calcium**. This substance undergoes an energetic chemical action with water — resulting in the formation of lime, and the gaseous hydrocarbon, *acetylene*. The action is chemically represented by the following equation:—

$$\underset{\text{Calcium Carbide}}{CaC_2} + \underset{\text{Water}}{H_2O} = \underset{\text{Lime}}{CaO} + \underset{\text{Acetylene Gas}}{C_2H_2}$$

It is this gas which has claimed the public attention for the last few years as an illuminant. Compared with the ordinary coal-gas of 16 candles, acetylene shows an illuminating power of 240 candles; in other words, it is fifteen times as rich in light-giving power.

It is only of late, however, that acetylene has come into use, owing to the difficulty in procuring a supply of calcium carbide. Plant for the extensive manufacture of the material has been erected at the Falls of Foyer in Scotland, and great care is exercised over the purity of the lime and coke-dust; the product is a hard, tough, metallic-looking substance, of a dark-gray colour, tinged with purple. It is produced very nearly chemically pure, which is an important feature, as impure samples are dangerously explosive when producing gas under cover.

There are a number of firms throughout the country who supply **apparatus** for the generation of acetylene from calcium carbide; all are of different design and construction, but mainly they rely on two methods of producing the gas. The first practice is the use of a small gasholder, with the usual water seal, but arranged internally for the attachment of a perforated trough or wire basket containing the carbide of calcium. As the gas is consumed, the holder and basket fall to the water-level. The water acting on the carbide readily, acetylene is given off in copious volumes, and the holder rises again owing to the accumulation of gas. Upon the rising of the holder, the carbide is removed from contact with water, and the gas-producing action ceases.

The second type of generator is shown in fig. 665, and consists of a reservoir, having two or more chambers connected by pipes or small openings at the bottom. A quantity of water is supplied to one chamber, and carbide of calcium placed in the other. Communication between the two chambers is established, the water reaches the carbide, chemical action ensues, and acetylene gas is produced. Finding no outlet, however, the gas begins to exert a pressure, which, acting upon the surface of the water, forces it through the channels back into the water-chamber. The generating action consequently ceases so long as the pressure is maintained. Thus the supply of gas is automatic and constant, fall of pressure admitting water and permitting the production of gas, and a rising pressure ending the generation of the gas by compelling the water to leave the chamber. In this latter method a good system of governing the pressures of the acetylene gas is a necessary concomitant.

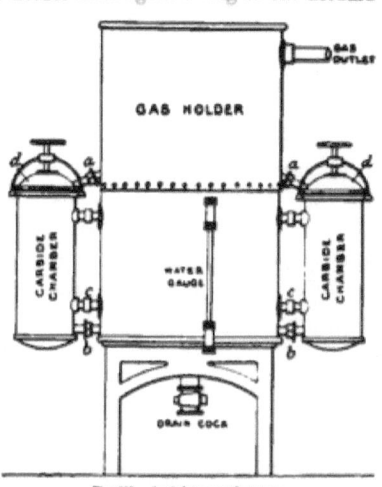

Fig. 665 — Acetylene gas Generator.

a a, gas inlets to holder; b b, water inlets to calcium-carbide chambers; c c, supports for calcium-carbide chambers; d d, removable chamber covers.

In either system, **the mode of obtaining the gas is** remarkably simple, the apparatus can be managed and worked by any intelligent boy or servant. It is a very clean method of procuring light; and the **apparatus** is compact and cheap. Many instruments for generating the gas can be charged for making large quantities continuously.

The light obtained from acetylene is a very brilliant white light, the gas being burnt in flat-flame burners with very diminutive orifices. It is distinguished from coal-gas by the great absence of any internal blue flame, such as is usually noticed near a gas-burner. An acetylene flame, with a superficial area equal to that of a shilling, will give an illuminating power of 20 to 25 candles. Such an intense light is not congenial to the eyes, and necessitates the use of globes and shades.

Hitherto some difficulty has been experienced in providing a **suitable burner for acetylene**; even with those having exceptionally small orifices, the gas will deposit carbon gradually, till the burner is rendered useless. A device of considerable ingenuity has brought an effective burner into the market. In the steatite tip, about one-eighth of an inch below the orifice for the gas, four small holes are drilled radially into the gas-way of the burner. By this means a small quantity of air finds its way into the gas before ignition, being drawn into the gas-channel by the efflux of the gas. The ignition in consequence takes place at a little distance from the extremity of the steatite tip. The value of the burner described is further enhanced by the use of two jets, placed at an angle that causes the two flames to impinge on each other, tending as it does to produce a steady broad light.

The question of illuminating country houses by acetylene deserves serious consideration. The facility with which the gas can be made, the reduced size of plant required for its generation, and the brilliancy of the product, naturally recommend the system to every investigator of its merits.

On the other hand, there are **two most essential conditions** required with the use of this illuminant. The first consists of a very dry and well-ventilated store for the carbide of calcium, as this readily absorbs moisture from the atmosphere and gives off acetylene gas. For this reason, an apparatus which provides for the attachment of charges of carbide in cylindrical vessels to the sides of the generator, as shown in fig. 662, has an advantage, owing to a large quantity being securely imprisoned, yet ready for work immediately it is required.

The second condition referred to, and which should be a *conditio sine quâ non* in regard to the adoption of the system, is **the absolute soundness of the gas-fittings** inside the house. An escape of acetylene gas is fraught with grave consequences. Even in small quantities its effects have proved serious; and every effort should be made to obtain perfectly-sound fittings. Acetylene gas, when inhaled, acts injuriously upon the hæmoglobin of the blood, and consequently is a direct enemy to the human economy.[1]

---

[1] There is also a serious danger of explosion when this gas is used. Any mixture of air and acetylene, in which the latter is within the limits of 3 and 82 per cent of the former, is very explosive. So grave is the danger, that in

Carbide of calcium can be obtained from the Acetylene Illuminating Gas Co., at the Falls of Foyer, at a price of £20 per ton. This company has the legal monopoly of making the substance by electrical furnaces, in the United Kingdom. One ton of carbide will produce practically 11,000 cubic feet of gas; 1000 cubic feet of gas will consequently cost £1, 13s. 4d. Burners consuming ½ cubic foot per hour will give a light of 25 candles, and few makers care to go above this illuminating power, owing to the uninviting colour of the flames.

CHAPTER IV.

SPIRIT-GAS.

There is another system of producing illuminating gas, which, though not used extensively, is still in vogue. The gas is obtained by the complete **evaporation of suitable spirits.** Such fluids are obtained from crude oils by fractional distillation. They are highly inflammable bodies, and possess the

---

Germany the use of the gas was at one time entirely prohibited, and in our own country an Order in Council, dated February 26, 1897, was made, prohibiting the keeping of carbide of calcium without a license. This stringent Order was somewhat relaxed in the following July, the Order now in force running thus:—

"The quantity of carbide of calcium which may be kept without a license shall be as follows:—

(a) Where it is kept in separate substantial hermetically-closed vessels containing not more than 1 lb. each, ... 5 lbs.
(b) Where it is kept otherwise, ... None."

The necessity for these regulations is manifest when we remember that moisture, acting on the carbide of calcium, liberates acetylene, which, when mixed with air in certain proportions, provides a highly explosive compound.

A more recent Order in Council (dated November 26, 1897) deals with liquid or compressed acetylene, and is so important that it must be quoted almost in extenso:—

"Whereas acetylene when liquid or subject to a certain degree of compression is specially dangerous to life or property by reason of its explosive properties.

Now, therefore, Her Majesty is pleased by and with the advice of Her Privy Council to order and declare, and be it ordered and declared as follows:—

Acetylene when liquid or when subject to a pressure above that of the atmosphere capable of supporting a column of water exceeding one hundred inches in height and whether or not in admixture with other substances, shall be deemed to be an explosive within the meaning of the said Act, subject to the following exception: that if it be shown to the satisfaction of the Secretary of State that acetylene, declared to be explosive by this Order when in admixture with any substance, or in any form or condition, is not possessed of explosive properties, the Secretary of State may by Order exempt such acetylene from being deemed to be an explosive within the said Act.

And whereas by section 43 of the Explosives Act, 1875, it is provided that Her Majesty from time to time by Order in Council, may prohibit, either absolutely or except in pursuance of a license of the Secretary of State under the said Act, or may subject to conditions or restrictions the manufacture, keeping, importation from any place out of the United Kingdom, conveyance, and sale, or any of them, of any explosive which is of so dangerous a character that in the judgment of Her Majesty it is expedient for the public safety to make such Order.

And whereas it is in the judgment of Her Majesty expedient for the public safety that acetylene, when an explosive within the meaning of this Order, shall be prohibited.

Now, therefore, in pursuance of the above-mentioned provision of this Act, Her Majesty is pleased, by and with the advice of Her Privy Council, to order and prescribe that acetylene declared to be an explosive by this Order shall be prohibited from being manufactured, imported, kept, conveyed, or sold."—Etc.

property of being entirely converted into gas. They have a very low flashing point, many degrees below that of freezing point, and are readily volatile at ordinary temperatures. The two principal kinds of petroleum spirit used for the purpose are gasoline and carburine. The former has a specific gravity of ·660, and the latter of ·680. The liquid is invariably stored in the open air. As gasoline requires another step in its preparation beyond carburine, and is consequently more expensive, the latter (or ·680 spirit) is most frequently used.

The apparatus used for the conversion of this liquid into gas for illuminating purposes is very compact, consisting of a small gasholder 3 feet 6 inches in diameter, capable of holding from six to eight cubic feet of gas. The columns of this gasholder, carrying the guide-runners, support a platform, on which a diminutive retort is placed. The retort is heated by atmospheric burners. A pipe leading from the spirit-tank, which is fixed at a higher level, enters the retort, the supply of liquid being regulated by a valve of an ingenious design and careful workmanship. The outlet from the retort is placed at the top, and is connected to the inlet of the gasholder below. Upon starting the apparatus, the supply of gas is automatic. The burners under the retort are ignited, and a temperature between 400° and 500° Fahr. imparted to the vessel. The cock admitting the liquid from the tank is opened, and the supply of gas is controlled by the valve. The valve-lever is connected to the gasholder, and the rising or falling of the holder actuating the valve decreases or increases the flow of spirit into the retort. The carburine, coming in contact with the heated surface of the retort, is immediately and completely vaporized into a rich illuminating gas, which, by the pressure created, passes into the holder.

The rapidity with which the supply of gas can be obtained, and the readiness with which the process will adapt itself to the consumption of one or three hundred lights, have procured for this system **a considerable amount of patronage**. The gas, however, although of a rich illuminating power (being of some 60 to 70 candles in value), is expensive, costing from 6s. to 7s. per thousand cubic feet, according to carriage, which is considerable with this class of inflammable material.

Of all the methods described for producing illuminating gases, **this system demands the least attention** or cleaning. The spirit is entirely evaporated into gas, and leaves no deposit in the holder or retort. It is quickly brought into action, and as quickly stopped. It will produce a sufficient quantity for 300 lights, or for maintaining only the two atmospheric burners necessary for heating the retort. The retort and gasholder can be placed in the cellar of the house without danger or nuisance. Ordinary oil-burners, or mushroom regenerative burners, may be used for its consumption.

SECTION XV.

THE SANITARY ASPECT OF DECORATION AND FURNITURE

BY

EDWARD F. WILLOUGHBY, M.D. (LOND.)

DIPLOMATE IN STATE MEDICINE OF LONDON UNIVERSITY, AND IN PUBLIC HEALTH OF CAMBRIDGE UNIVERSITY
AUTHOR OF "PUBLIC HEALTH AND DEMOGRAPHY," "HEALTH OFFICER'S POCKET BOOK," ETC.

SECTION XV.

THE SANITARY ASPECT OF DECORATION AND FURNITURE.

1. *GENERAL CONSIDERATIONS.*

The essential purposes or **aims** of the internal arrangements, decoration, and furniture of our dwellings are health and comfort, together with the pleasure derived from the contemplation of beauty of form and colour, which may be itself made conducive to health and comfort or the reverse. The necessary conditions of health are air, light, warmth, and the absence of damp and dirt; and no artistic effects should or need be allowed to conflict with these primary and paramount considerations.

Light is scarcely less necessary to animal life, or, at any rate, health, than it is to plants. The production of hæmoglobin, the red pigment of the blood of animals, is dependent on the enjoyment of the fullest possible amount of light, as is seen when one compares the ruddy complexion of persons who are constantly in the open air, with the pallor of the denizens of dark dwellings, or of those who pass most of their time indoors, though the freer oxidation of the blood by the respiration of a purer air is a no less active factor. Still, the influence of sunshine—that is, of the maximum amount of light—is scarcely less important as an auxiliary to oxidation of the blood.

Even in the house, light, and light of particular colours, has a powerful influence on the senses and the mind through the medium of the nerves. Dull, dark, dingy colours are depressing to the spirits, glaring reds and yellows are more or less irritating, while white light is healthful, and the chemical rays, blue and violet, are agreeable and soothing. **Persons in robust** health may not be much affected by these differences, but the **invalid and sensitive are** very susceptible to the influence of colour, which is doubtless no less real, though less in degree, on all.

Dirt was defined by Lord Palmerston as "matter in the wrong place", and this description can scarcely be improved on. The grosser and more repulsive forms of dirt find no place, or lasting lodgment, in a well-regulated household, where the refinements of civilized life are rightly appreciated; but there are other forms that are ubiquitous, and elude all ordinary efforts made for their prevention and removal.

Foremost among these is dust, or the fine particles of organic and inorganic matter which, floating in the air, are deposited on every horizontal surface and adhere to every inequality on walls and furniture. Dust enters our houses with the fresh air from without, and is formed within by the wear and tear of our carpets, the combustion of fuel in our fire-grates, &c. It is of the most varied composition. That from without consists of fine particles of sand and other earthy matters raised by the wind, and produced in enormous quantities in dry weather, especially by the traffic in roads and streets, together with dried and pulverized horse-dung, the proportion of which varies with the efficiency of the scavenging. In towns, especially in dull or foggy weather, the air contains a large amount of soot in finely-suspended particles, or in floating masses or "blacks". The internal dust is composed of the light ash from coal-fires, with more or less carbon of smoke, the debris of carpets worn by treading and in sweeping, and that of clothing, &c., with epithelial scales shed from the skin, fine hairs, &c. Dust is deposited everywhere; it permeates carpets, rugs, mats, the covers of furniture, and the interior of cushions; it clings to curtains and the margins of books; adheres firmly to rough wall-surfaces, especially flock-papers; fills the crevices of the flooring, and accumulates in the space beneath the boards.

Moisture within our houses is not wholly due to external conditions; much of it is produced within the house itself by the combustion of gas and oil, and by the lungs of the occupants. These are the chief sources of the water, which, condensed by the cold glass of the windows, runs down in streams on a winter's night, or, freezing, encrusts the panes with feathery ice. One frequently hears it urged against impervious wall-surfaces, that a porous wall affords a certain amount of ventilation. But, true as this is of the bamboo-mat walls of an Indian bungalow, the quantity of air that can pass through brick and plaster is inappreciable, and the evils consequent on the absorption of moisture from without and its evaporation within, are so real and grave, that the contention is unworthy of serious consideration. The notion that impervious wall-surfaces are themselves a cause of damp in rooms,—a notion based on the fact that drops of water may be seen standing on or running down such walls,—is the result of a misapprehension of the source of the moisture, and the cause of its condensation, which are the same as those of the water on the windows; and it would be as reasonable to maintain that these should be made of muslin or paper instead of glass. The surface of a plaster or papered wall always appears dry, simply because the moisture, instead of being condensed, is absorbed, to be again evaporated when the room is warmed, leaving behind the organic matters exhaled from the lungs, until in course of time the wall becomes saturated with decomposing animal matter, even to the extent of emitting a musty or offensive odour.

2. *WALLS AND CEILINGS.*

Limewashing is suitable only for stables, cowsheds, fowl-houses, and outbuildings. It is made by mixing freshly-burnt quicklime with water to the consistence of cream; no size is used, and it therefore easily comes off, but its value consists in its being a powerful insecticide and germicide, second only to corrosive sublimate as a disinfectant. It should be reapplied at regular periods.

Whitewashing, although in some districts the term is applied to limewashing, is in reality a totally different operation. "Whiting", *i.e.* finely-ground chalk, which has no disinfectant properties, is mixed with water, together with size and alum to prevent its rubbing off. It is used for ceilings and rougher walls, and may be tinted by the addition of some colouring material. It is, in fact, a form of distemper.

Distemper is the name given to all colouring processes in which the pigments are mixed with size, in distinction from painting in oils, in which the vehicles are boiled linseed-oil and turpentine. Painting in distemper is ill adapted for woodwork, to which it does not adhere well, and is practically limited to plaster, which, however, is in itself too absorbent, and requires a previous dressing with a coat of whitewash, prepared with a larger proportion of size, some alum, and soft soap. The objection to preparing the wall with a couple of coats of oil-paint is that the condensed moisture will cause the distemper to "run", and shows that distemper, unlike paint, is not impervious, but as absorbent as plaster or whitewash. The pigments employed, as prussian blue, indigo, ochre, umber, venetian red, &c., are for the most part perfectly harmless, but poisonous colours are sometimes used, as emerald green, which is arsenical.[1] And this caution should always be borne in mind, since distemper is not so firmly adherent as oil-paint, and there is therefore a risk of the pigment becoming detached and floating in the air as dust. If desired, it may be ornamented by panels or other designs laid on in different colours by means of stencil-plates, which may be had in every kind of pattern.

[1] A good green may be obtained by mixing prussian or indigo blue with yellow ochre.

There are, besides, several kinds of so-called **washable distempers**,—such as Orr's Duresco, &c.,—non-absorbent and durable, by means of which plaster walls may be made impervious to moisture. For this purpose, however, none of them can compare with the German water-glass.

Water-glass is much used in Germany for architectural purposes and decorative art. It is a silicate of potash, which, if once allowed to dry, cannot be again dissolved save by superheated steam or very prolonged boiling, but in the gelatinous form, like size, is easily soluble in hot water. Applied as a wash to the soft stone used for architectural decorative work and the tracery of windows, it preserves it from the disintegrating action of the weather, almost without altering its colour or appearance; while for inner wall-surfaces, it constitutes a non-absorbent and therefore washable coat perfectly impervious to damp. Like distemper, it may be employed as a medium or vehicle for colour, and if mineral pigments only are used, and it is laid on a freshly-prepared surface of plaster or cement, it sinks in, producing a perfect fresco painting. The great works of Kaulbach on the walls of the museum at Berlin, and of the Pinakothek at Munich, are frescoes in water-glass, and imperishable. If the glass be dissolved with little water and used hot, the effect closely resembles that of encaustic tiles or enamel. The various applications of water-glass, alike for sanitary and decorative purposes, deserve more attention than they have yet received. There is also a soda silicate, more easily obtainable in this country, and convenient for some purposes, but from its far greater solubility little superior to ordinary distemper.

Most of the mural paintings in old churches, &c., though commonly called frescoes, are really paintings in distemper, perhaps treated with a dressing of wax for their better preservation. True frescoes are rare; in them the colours, necessarily mineral, are ground fine, mixed with water or milk of lime, and laid on a freshly-prepared surface of plaster, with which they become incorporated while it is still moist. Only a small surface can thus be painted at a time, and once finished no subsequent touching-up or alteration is possible. Unlike distemper, the work is practically imperishable.

Oil-painting is the very best covering for walls of halls, dining-rooms, libraries, and indeed of all others. It is, however, rarely used for bedrooms, for which it would seem, on sanitary grounds, to be specially adapted, since in the event of infectious diseases occurring in the family, no stripping of the walls will be required, washing with corrosive sublimate, or even with soap and water, being sufficient for complete disinfection. Paint is absolutely impervious and non-absorbent, and easily washed. There is no danger to be feared from the use of poisonous pigments, since, unlike distemper, paint does not rub off.

The use of coloured wall-papers, though known in China from time immemorial, probably had its origin in Europe in the endeavour to find a cheap substitute for the costly tapestries which in the middle ages had taken the place of the mural paintings of the Greeks and Romans, or for the painted canvas hangings which were in vogue from the fifteenth to the seventeenth centuries. But though they have long been almost universally employed, they are, with a few exceptions of recent introduction, open to serious objections. Absorbent of moisture, retentive of dust and dirt, to say nothing of occasional infection, and (unless varnished) incapable of being washed, common wall-papers are in the highest degree insanitary. The "flock-papers", in which shoddy dust is made to adhere more or less to the surface of patterns printed in glue, are undoubtedly the worst; happily they have of late years gone much out of favour. The pigments used are of various kinds, many of the mineral colours being poisonous; but if the surface be glazed or varnished, there is little danger to be feared from preparations of lead, cadmium, &c. It is, however, quite otherwise with the volatile compounds of mercury, antimony, and arsenic, the last being by far the most hurtful, and, formerly at any rate, most extensively employed.

It is a popular but erroneous belief that **arsenic** is present in greens only; but while every shade of green can be had without arsenic, there is no colour— red, blue, brown, or white—which can as such be considered safe. The "simple" test, recommended in popular works, of the deep-blue colour brought out by the application of strong liquid ammonia, is utterly untrustworthy, being properly a test for copper, not for arsenic, and available in the case only of colours containing Scheele's green or arsenite of copper. Papers of inferior quality, to which the colour adheres but loosely, are the most dangerous; those, in fact, that are most used for bedrooms and the poorer class of houses. When acting as secretary to a committee on poisonous pigments, appointed by the National Health Society, with the aim of obtaining legislation such as had been long in force in Germany, Austria, Sweden, Denmark, and Holland, I had ample opportunities of studying the question. I have a book of samples selected from among 700 examined, which are arranged in pairs of similar colours and not unlike patterns, so that they could in every case be substituted for **one another to** please any **fancy**, one of each pair being **arsenical**, often highly so, and the other free from the least trace. Every colour is there represented. Those made by W. Woollams & Co. were all absolutely free; but **many** others, though guaranteed as such, proved to be more or less arsenical.

No doubt **many alleged cases of illness** thus caused will not bear scientific investigation, but others are incontrovertible. Though the effects are frequently obscure, dyspepsia and general derangement of the digestive and nervous systems, irritation of the mucous membrane of the eyes, and other like symptoms, relieved by change of residence and recurring on return, should be regarded as highly suspicious. Some persons, it should be borne in mind, are much more susceptible than others.

Arsenic, though absent from the paper, may be present in the paste, being sometimes used as a preservative to keep it from going sour or putrid. When a fresh paper is to be put on, the wall should be thoroughly stripped, for the slovenly practice of leaving coat over coat of dirty papers and foul paste is, to say the least, nasty and unwholesome.

Cretonnes and imitation Indian muslins are frequently arsenical, some having been found to contain as much as 20 grains in the square yard. Aniline dyes, more used for textile fabrics than for papers, are often prepared with arsenic, which is removed imperfectly or not at all. In Germany, where the best aniline colours are made, the use of arsenic in their manufacture is illegal; and though, in view of French competition, an exception was made in favour of the export trade, the use of arsenic was beset by such vexatious restrictions, that the manufacturers were not long in finding equally energetic, though innocent, reducing agents; but no such control is exerted in this country or in France.

Of late years a large number of new materials intended as **substitutes for wall-papers**, and possessing various advantages over them, have been placed on the market. Some are highly artistic, but too costly for the smaller class of houses; others are little dearer, and from their greater durability cheaper in the end, than papers of fairly good quality, and all are washable. Muraline, the least-removed in appearance from ordinary paper, is inexpensive, very durable, and resists scrubbing with soap, its speciality consisting in the preparation and laying on of the colours, which are not mixed with size and water. Fisher's permanent wall-hangings hold an intermediate position between this and the class of hard, stout, embossed materials, such as Tynecastle canvas, Anaglypta, and the Salamander decorations,—the last, being made with asbestos, is uninflammable. All these require a coat of varnish, or a special preparation adapted to each. Though richly embossed, the fact that they can be washed with impunity obviates the objection incident to ordinary papers with raised patterns, that of giving a lodgment to dust. But where cost is not a consideration, nothing can approach Lincrusta-Walton as a non-absorbent, extremely durable, and highly artistic wall-covering. The designs, in the best style of decorative art, stand out in bold relief, but being solid and made of an elastic material, are almost insusceptible of injury. It is attached to the wall by a special adhesive and waterproof "glue", a preparation of caoutchouc, and may be gilded, stained, or painted in one or more colours before being varnished. It is made in lengths, panels, and borders, suited for walls, dadoes, the faces of pilasters, friezes, &c., and when deeply stained presents a close imitation of black carved oak, yet is otherwise sufficiently characteristic to escape the stigma of sham.

Ceilings are commonly whitewashed for the better reflection of the light, but the effect while fresh is not pleasing, and they soon become dingy and black from smoke. Tinting, with picking out of the cornices with colours in distemper, obviates this to some extent and gives a more agreeable effect; and papers are sold for the same purpose, but the objections to absorbent wall-surfaces apply equally to ceilings, though they are not exposed to friction. There is no conceivable reason why they should not be made impervious and washable by paint applied to the plaster or on canvas, if it can be made to adhere firmly. The best effect is obtained by panels of Lincrusta, or of those materials which most nearly resemble it. But the practice, adopted by some architects of parsonages and quasi-ecclesiastical or collegiate houses, of dispensing with lath and plaster, and ceiling the rooms with boards nailed to the joists and varnished, might well be more generally followed, especially in halls, dining-rooms, and libraries.

3. *BLINDS AND CURTAINS.*

All **curtains** are unavoidably collectors of dust, but this objection may be minimized by their being hung simply on poles without the needless and dangerous addition of heavy and insecurely-fixed cornices with heavy fringes, which it is impossible to dust; while in drawing and morning rooms, nothing can surpass for convenience, elegance, and cleanliness, the pretty Indian curtains made by stringing short pieces of bamboo.

Venetian blinds are objectionable, as they collect dust, which is removed with difficulty, and are complicated and always getting out of order. Florentine blinds, as they are called, are preferable for excluding the direct rays of the sun without interfering with the air and view, or still better, the Helios, consisting of several broad canvas louvres adjustable at any angle by a crank arrangement. The "tropical revolving blinds", devised by Mr. D. Radclyffe for shading plant-houses, are equally **applicable** to domestic use, and, like the Indian matting blinds which they nearly resemble, give a soft light and free passage to the air, permitting a good view from within, but not from without, thus securing the same privacy as if they were opaque.

4. WOODWORK

The practice of covering all woodwork with paint, concealing the natural beauty of the grain, and perhaps substituting some vulgar imitation, is at once senseless and useless. The preservation of the wood, and facilities for keeping the surface clean, are equally attained by the use of varnish or polish, which heighten the colour and vein, and bring out the natural characters of such woods as oak, mahogany, walnut, rosewood, maple, and satin-wood; and even the common pitch pine, red deal, and white deal are, when so treated, by no means devoid of a beauty, which if desired may be enhanced or varied by staining. The only drawback is the frequent occurrence in the inferior kinds of timber of knots, which it cannot be denied are unsightly; but it is quite possible to select boards of sufficient size for most purposes free from these disfigurements, and there are many American woods of the greatest beauty, which can be obtained in any lengths free from the slightest flaw or defect, as anyone may observe in the interiors of the London tramcars. I do not deny that rich effects may be obtained by a judicious contrast of colours or shades in painting, but the blank white which shows up dirt, wears thin and bald by frequent use of soap, and in the sulphur-charged air of towns soon acquires a yellowish or brownish hue by the changes effected in the white-lead, is utterly tasteless, while graining is a mere sham, and, as such, a violation of the elementary principles of art.

Varnish over the natural shades of the wood does not show up every touch of the hands or clothing; it undergoes no chemical action, and is easily cleaned by a cloth with water alone, while new coats may be applied as often as may be necessary with little trouble.

5. FLOORS AND FLOOR-COVERINGS.

The bare and ill-fitting boards until recently used for flooring in all but the very best class of houses, with their original inequalities exaggerated by subsequent shrinkage, necessitated the covering of the entire floor with carpets, fastened down, and removed but once or twice in the year. Such carpets become permeated by dirt, which accumulates beneath them, while the act of sweeping produces large quantities of dust, and is, indeed, almost the sole cause of flues, which are composed chiefly of the detached fibres of wool. On every ground of health, comfort, and appearance, it is desirable that the floors of all rooms alike should be made perfectly smooth and free from chinks, then stained and rendered as nearly non-absorbent as possible.

Varnish soon wears away, and sometimes assumes a dull whitish hue; far better (though somewhat more labour and expense are involved in the first application) is bees'-wax melted with a little boiled linseed-oil and turpentine, and rubbed on with a flannel. While varnish requires a previous use of size, the best preparation for wax is the repeated saturation of the boards with boiled linseed-oil, which, hardening, renders them almost impermeable. If the floor is old, and more or less charged with soap and grease, it should be well planed before any attempt at staining. The stain should be one that will soak deep into the wood, and I know of none better than a strong solution of permanganate of potash, which has the advantage of not merely permeating, but of entering into a chemical combination with the structure of the wood. It costs about half a crown a pound, and, dissolved in water, will impart to common deal any shade from light-brown to black oak, according to the strength of the solution.

With such floors, **loose squares of Turkey carpet,** Persian or Kirghiz rugs, or imitations of the same, may be laid wherever needed, the sides and less-frequented parts of the floor being left uncovered. Sweeping and scrubbing are alike unnecessary; the loose carpets and rugs are rolled up and taken out of doors to be shaken, while the boards need only to be wiped over with damp cloths. In bedrooms, the sanitary advantages of this arrangement are obvious, especially since on the earliest suspicion of infectious disease, the rugs can be removed without any trouble, and the cost of subsequent disinfection avoided.

Floorcloth and linoleum are alike washable, and do not give rise to dust, though it may accumulate beneath them unless they are glued down to the floor. Cork-carpet is a somewhat similar material, but more absorbent, and also less noisy.

6. *FURNITURE.*

Within the last ten years there has been a most gratifying improvement in **the popular taste as regards furniture.** In drawing-rooms, however, there has been a tendency to overcrowding with the trivial and the useless. Especially has this deterioration been manifested in the decoration of the walls of drawing-rooms, boudoirs, and the like. That the art of Japan has had an influence for good in substituting simplicity of design and graceful forms for floridity and tasteless elaboration, admits of no dispute; but the imitation has been pursued with more zeal than knowledge. The Japanese do not smother their walls with fans and plates, and their instinctive sense of the fitness of things would be shocked at sham shoes or banjoes of silk and cardboard. Indeed the scantiness of the furniture and ornaments in a Japanese room surprises the European, no less than the elegance and charm of the little that he sees.

In furnishing a room, everything that interferes with the freest ventilation, that favours the accumulation of dust or hinders its removal, should be rigidly excluded. A desk, cabinet, sideboard, chest, or other heavy piece of furniture, which could not well be raised on legs, may, without detracting in the least from its solid and massive appearance, be made easily movable in every direction by being fitted with ball-rollers of gun-metal, which are quite invisible, and so incomparably superior to castors that the wonder is that they have not been long since applied to tables, heavy chairs, and couches. Leather covers of chairs and couches are better firmly stuffed and tightly stretched without "buttoning down", which, like all needless recesses and lodgments for dust and dirt (as deeply-embossed plushes, fluffy mats, and furs), are to be so far as possible avoided.

7. SWEEPING, DUSTING, &c.

Though the great aim of the architect and householder should be **the prevention of dust**, absolute success is unattainable, even in a house warmed and ventilated without open windows or coal fires in the rooms. This subject has already been touched upon in these volumes, particularly in Section II., Vol. I., and Section XII., Vol. II., so that one hint must suffice here, namely, that much dust will be avoided by fitting to all fire-grates movable bars and pans to receive the ashes.

Sweeping, as ordinarily done, contributes largely to the production of dust, and dusting, so-called, utterly fails to remove it. Flues and snippings, &c., may be taken up by a dust-pan and soft hand-brush, and carried out of the room, and carpets, which are nailed down, are best swept by the mechanical sweepers, which cause the minimum of friction and raise no dust. But loose carpets or rugs on polished floors have this advantage, that they may be rolled up and carried out of doors to be beaten or shaken, while the boards are wiped with damp cloths.

All mantel-pieces, ornaments, picture-frames, and the wood and leather of furniture, should be **wiped down with cloths**, wrung out in water so as to be just damp, and frequently renewed by dipping and wringing as fast as they become soiled. The quantity of dust thus removed is surprising, whereas dry dusting simply raises the dust into the air to be deposited again in the course of an hour or so. Non-absorbent walls and varnished or waxed floors, as well as tiles and natural or artificial stone, may be kept clean by wet cloths without soap, but bare boards must be scrubbed. In doing this, however, the brush should follow the grain, and not be moved in circles, as this is apt to give a smeared appearance to the wood when dry.

Furniture polish, composed of spirits, turpentine, and oil, with or without other ingredients, should be indulged in as sparingly as possible, for it acts as a solvent on all kinds of varnish, and simple but vigorous rubbing with a damp cloth is much to be preferred. Even on plain unvarnished mahogany, as dining-room table-tops, the shine given by "polishing oils" is transient and illusory.

Oil-paintings should never be touched with soap; gentle rubbing with a handful of cotton-wool, with a to-and-fro movement, first in one direction and then at right angles, not round and round, will generally suffice, though if very dirty a soft handkerchief, just moistened with water, may be used. Gold frames are best cleaned by rubbing with a cloth wrung out in the best paraffin oil, which will soon evaporate if the excess be removed by a clean dry cloth. So long as a painting is protected by varnish the colours are unaffected by the products of the combustion of gas, but should the paint be found after cleaning to be exposed, a fresh coat of mastic should be laid on as thin as possible, and any "bloom" that may appear on drying removed by gentle wiping with dry cotton-wool.

Gas is most destructive to the **leather bindings of books**, but they may be protected against its action, and if not too far gone, restored to a great extent, by the application of gold size, which is also very useful for renovating French-polished furniture.

8. *PLANTS IN ROOMS.*

As to the pleasure derived from the cultivation of **plants in rooms**,—though very few are capable of resisting the poisonous action of the products of the combustion of gas,—there can be no difference of opinion, but many persons labour under the delusion that their presence is injurious to health. It is true that certain powerful perfumes may induce headaches, but the notion with regard to plants in general is based on a misconception as to the nature of the nutritive process incorrectly called their respiration, which is assumed to be similar to that of animals, whereas it is in fact precisely the reverse. Animals absorb oxygen from the air, giving out in exchange an equal volume of carbonic acid, which plants absorb under the influence of light, converting it into starch and woody fibre **or cellular tissue**. They thus preserve the purity of the air by removing **the poisonous gas** evolved by animals and the combustion of hydrocarbons, and maintain the equilibrium of nature. This process is suspended **in the** dark, but **the amount of** oxygen they take up even then in true respiration is insignificant, and **when** one considers that far the greater part of the structure of a plant is built up of the carbon derived from the carbonic acid in the air, one cannot but feel that their power as purifiers of the atmosphere, which otherwise would in time become unfit for the maintenance of animal life, must be enormous.

There are, however, very few plants that can be grown in **rooms lighted by gas;** aspidistras, dracænas, and some palms, robust ferns, and cactuses, are the least susceptible. The compounds of sulphur, among the products of the combustion of coal-gas, are most fatal to vegetable life, so that in this respect the smokeless incandescent gas-light possesses no advantage.[1] Oil and candles, being free from sulphur, are harmless, and electric light, like that of day, is actually beneficial. Where gas is used, the difficulty may be overcome by window-cases, the inner sashes being closed so soon as the gas is lighted. Such cases fitted to windows, the outlook from which is not pleasant, may be made extremely ornamental, and if the aspect be northerly or the sunshine shut off by neighbouring buildings, they may be filled with ferns, palms, and other plants which prefer the shade.

Hermetically-sealed glass cases, or simple bell-glasses closely fitting the pan of earth beneath, and filled with delicate ferns, selaginellas, &c., constitute most elegant table decorations. In these "Wardian" cases, plants will thrive for years without any attention whatever, provided only they have as much light as the particular kinds require.

The use of foliage, interspersed with a few flowering plants, need by no means be limited to rooms and windows. A skilful florist can convert an area into a fernery and moss-clad grotto, or a backyard into a winter-garden. In these positions the choice of plants, and especially of flowers, is restricted to such as love the shade, but the possibilities of roofs are vast. On the warm sunny "leads" of a wing or addition, sheltered by the wall of the main building, a luxurious conservatory or orchid-house may be erected, approached by glass doors and curtains from a passage or landing; on the shadier side may be a fernery, while the roof of the house itself, if the style permit of its being flat, would afford an excellent position for a large conservatory and a terrace, with seats and boxes of plants, providing a garden where there would otherwise be none, and where a purer air could be enjoyed than in the streets or the confined gardens of houses in towns. In this respect we might well take a lesson from the peoples of Eastern and tropical countries, among whom the house-tops are the favourite evening resort of the family.

[1] Except that, in a given time, and for the same amount of light, it burns less gas than an ordinary burner, and consequently may be expected to have a less rapid effect on plants.—ED.

www.ingramcontent.com/pod-product-compliance
Lightning Source LLC
Chambersburg PA
CBHW032149160426
43197CB00008B/841